AF709195

Taking the Wheel

by Wendy Stein

Designed and illustrated by Elaine Fichera

EACH ONE TEACH ONE

New Readers Press
Publishing Division of Laubach Literacy International
1320 Jamesville Ave., Box 131, Syracuse, NY 13210

"EACH ONE TEACH ONE"

ISBN 0-88336-455-7

© 1978, 1986
New Readers Press
Publishing Division of Laubach Literacy International
Box 131, Syracuse, New York 13210

All rights reserved. No part of this book may be reproduced or transmitted in any form or by any means, electronic or mechanical, including photo-copying, recording, or by any information storage and retrieval system, without permission in writing from the publisher.

Printed in the United States of America

9 8 7 6 5 4 3 2

Cover: Designed by Caris Lester
Illustrated by Joe Glisson

A note on pronouns

For a long time, people have used only male pronouns *(he, his, him)* to refer to persons of either sex. But that is unfair to half the human population! So sometimes we will use female pronouns *(she, her, hers)* and sometimes we will use male pronouns to refer to persons of either sex.

Table of Contents

Part One

Safe driving means knowing yourself, your car, the rules of the road, and basic driving skills. That's what this book is all about. But, first, let's talk about what you want to know most of all — **getting your driver's license and . . .**

Getting Ready to Hit the Road

1 Getting and keeping your license

A driver's license! It's many things to many people. Freedom! Independence! New thrills! Status! A more exciting social life!

What does a license mean to you?

A driver's license can mean all the things above. But let's get down to basics. Getting a license simply means getting the right to drive. In every state you must have a driver's license if you want to drive on public roads.

And since you want to be a safe driver, getting a license means something more. It means being trusted with an important duty—to help keep our highways safe. That's why your state tries to be careful about who gets and keeps a driver's license.

To earn a license, you must pass a written test, a road (driving) test, and a vision test. By passing these tests, you show that you:

– know the rules of the road
– can handle the car
– can see well enough to drive safely.

Learner's permit

A learner's permit gives you the chance to practice driving before you take the road test. A learner's permit is called an instruction permit in some states.

With a learner's permit, you may drive on public roads. But you must have a licensed driver sitting next to you in the car.

To earn a learner's permit, you must pass a vision test. You also must pass a written test about traffic rules and signals.

Where do you apply for a learner's permit? Go to the local office of the department of motor vehicles. You must fill out a form there. You must bring proof of your age and your identity (who you are). If you are younger than the legal driving age, you need your parent's OK.

The eye test is very short and simple.

Rules for learner's permits vary from state to state. In some states, a person with a permit may not drive between sunset and sunrise. In some states, a person must be 16 years old to apply. In some states, a person who has finished a driver education course may apply sooner. Your state's driver's manual will tell you the rules for applying in your state. It will also give the rules for driving with a learner's permit.

Road test

The road test is the last test to pass before you earn a driver's license. You must be the legal driving age to take the test.

A tester from the department of motor vehicles will take you on the road test. The tester will not try to trick you. He or she will not ask you to do anything that's not legal.

The tester judges you on your attitude and your

handling of the car. She judges you on how well you follow traffic rules.

You are graded on:

- starting the car
- entering traffic
- turning
- backing up
- stopping
- parking, and
- driving in traffic.

You might make small mistakes and still pass the test. But you can't make too many small mistakes. Serious mistakes, such as driving too fast or steering poorly, will keep you from passing. You will flunk if you break a law or cause an accident.

Types of licenses

There are different kinds of driver's licenses. You need to pass extra tests for some of them.

1. **Operator's license.** This license allows you to drive a car, van, or light truck.

2. **Commercial operator's license.** This license allows you to drive a truck that carries goods.

3. **Chauffeur's license.** This license is for taxi drivers, school bus drivers, and chauffeurs. It is for persons who are paid to carry other persons in a vehicle.

4. **Motorcycle license.** With this license, you may ride a motorcycle on public roads. In some states, you must earn an operator's license before you can earn a motorcycle license. In other states, an operator's license is not needed.

Once you earn your license . . .

Carry your license with you whenever you drive. Keep it in your wallet. Don't leave it in the glove box of the car. Don't leave it lying around. Someone could steal it.

Your driver's license is proof that you may drive a car. It is also a good identification card. If someone stole it, he could pretend to be you.

Don't lend your license to anyone. You could be arrested for lending it.

You must renew your license every few years. The number of years varies from state to state. In many states, your license expires on your birthday every few years. The department of motor vehicles may send you a letter to remind you to renew your license. You may renew by mail in some states.

If you move to another state, you must get a new license from that state. When you get a license from your new state, you give up your old license. You cannot have two licenses.

You should apply for a new license within about 60 days after you move. The time period may be different in your state.

You probably won't have to take a road test. You will have to pass a vision test and a written test. (In some states, persons older than 65 must take a road test, too.)

If you move to a new address in the same state, you must tell the department of motor vehicles. They need to know your new address.

It's yours to keep if . . .

A driver's license is yours to keep if you keep your state's rules for drivers. But what happens if a person doesn't keep the rules?

In some states, traffic violations (except parking) are reported to the department of motor vehicles. The violation goes on the driver's record. A certain number of points is given for each

The road test is finished! And the road tester shows that the driver did well on the test.

PORTION OF WRITTEN NEW YORK STATE DRIVERS LICENSE EXAMINATION

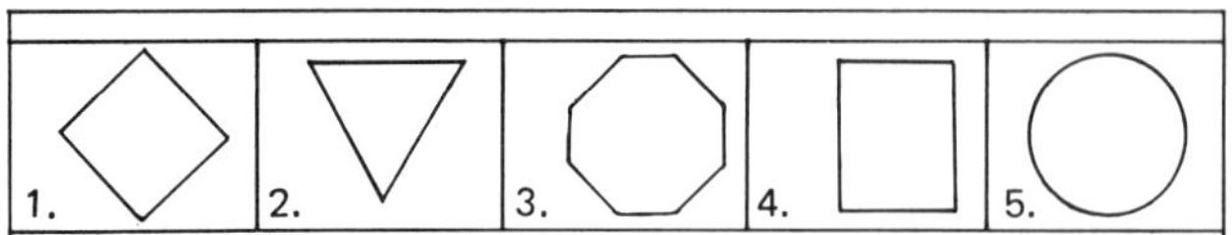

NO. 1

FIGURE NUMBER 1 SHOWN ABOVE IS THE SHAPE OF THE

A ROAD HAZARD SIGN
B SPEED LIMIT SIGN
C YIELD SIGN
D RAILROAD CROSSING SIGN

ANS.

NO. 2

FIGURE NUMBER 4 SHOWN ABOVE IS THE SHAPE OF THE

A RAILROAD CROSSING SIGN
B SPEED LIMIT SIGN
C YIELD SIGN
D STOP SIGN

ANS.

NO. 3

WHAT ARE THE COLORS OF A "STOP" SIGN

A WHITE LETTERS ON A GREEN BACKGROUND
B BLACK LETTERS ON A YELLOW BACKGROUND
C WHITE LETTERS ON A RED BACKGROUND
D YELLOW LETTERS ON A BLACK BACKGROUND

ANS.

NO. 4

YOU HAVE THE RIGHT-OF-WAY WHEN YOU ARE

A ENTERING A TRAFFIC CIRCLE
B LEAVING A PARKING SPACE
C ALREADY IN A TRAFFIC CIRCLE
D BACKING OUT A PRIVATE DRIVEWAY

ANS.

violation. If a driver gets too many points in a three-year period, he could lose his license.

You can lose your license with only one violation if it's a major violation. You need only one drunk-driving arrest to lose your license. Causing a fatal accident can be another reason for losing your license.

Violations from other states are counted, too, if the violations are reported.

A driver who has too many points is given a chance to defend himself. He can tell why he thinks his license should not be taken away. A state official then decides about the case.

National Driver Register

Mr. Winston was found guilty of drunk driving. His drunk driving had killed someone. His state revoked his driver's license. A month later, Winston moved to another state. He applied for a driver's license. He figured that his new state wouldn't find out about his record.

But he figured wrong. Both states take part in the National Driver Register program. Winston's first state reported to the National Driver Register in Washington that the license had been revoked. His new state wrote to Washington. It found out that there was a record of Winston's license being revoked.

The Register keeps track of people who have lost their licenses because of drunk driving. People who have broken a traffic law and caused a death are also in the Register.

Before this program began, a bad driver who lost his license could get a new license in another state. Now, a state can check the driving record of persons applying for new licenses.

Driver License Compact

Many states have made an agreement with other states. It is called the Driver License Compact. Suppose you get a speeding ticket when you are in another state. The state in which you were arrested reports that ticket to your home state. Then your home state has a record of your tickets, both at home and out-of-state.

Driver education

Driving can be very scary for a new driver. It can also be scary for the person trying to teach the new driver. This may be especially true if the teacher owns the car and is untrained to teach driving.

The beginning driver and untrained teacher may also be related to each other. This may make matters even worse.

Tempers flare easily. As tempers flare, the new driver makes more mistakes. As he makes more mistakes, the "teacher" gets angrier.

That doesn't sound like a very good way to learn to drive, does it?

That's why there are driver education courses. If you can't find a friend or relative who is a good teacher, try a driver education course. The courses are taught by professional driving teachers. They teach in public schools. Or they run their own private driving schools.

They teach how to handle the car safely. They try to help you gain a good attitude toward the car, driving, and other drivers.

There is **classroom training**. In the classroom, you learn the rules of the road and safety. Many classrooms have a machine with controls like those on a car. You *pretend* to drive in true-to-life driving situations. You practice on the controls before ever driving a real car.

Some states say all new drivers must have classroom training before they can get a license.

Another part of driver education is **behind-the-wheel training.** You may start out in a car with dual controls. There is a brake pedal on the passenger side. The teacher can stop the car if he has to.

Driving lessons are part of high school courses. They cost extra at private driving schools. Driving lessons are not required by states. But it is a good idea to take them.

Take your driving lessons from a pro, if you can. You won't have to put up with a cranky friend or relative yelling at you. You won't learn the bad habits of your friend or relative. You will learn the right way to drive—the safe way.

If you can't afford a pro, look for a friend with a good driving record. Avoid those who have had tickets or accidents.

Some studies say that driver education courses turn out the best drivers. The studies say these drivers have fewer accidents than persons who did not take driver education.

You won't get far in one of these car-like machines. But you'll learn how to use the controls.

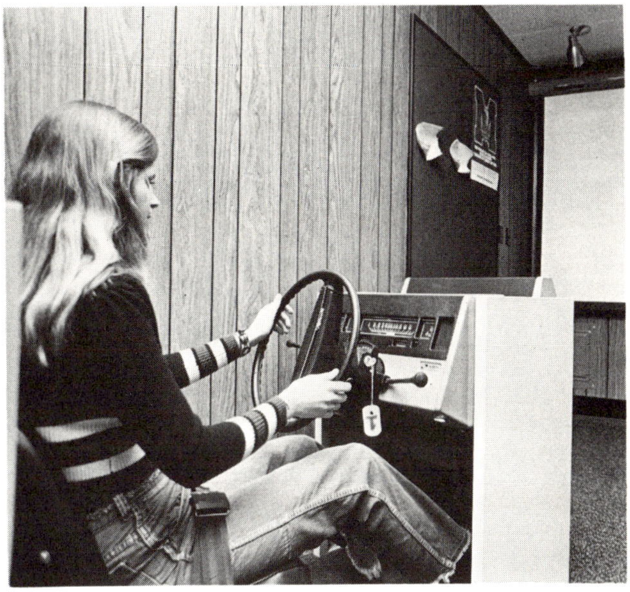

Other studies disagree. But insurance companies tend to value driver education courses. The insurance companies offer discounts to persons who have finished a driver education course.

Insurance payments are very high for new drivers, especially for young men. But driver education helps to cut the cost.

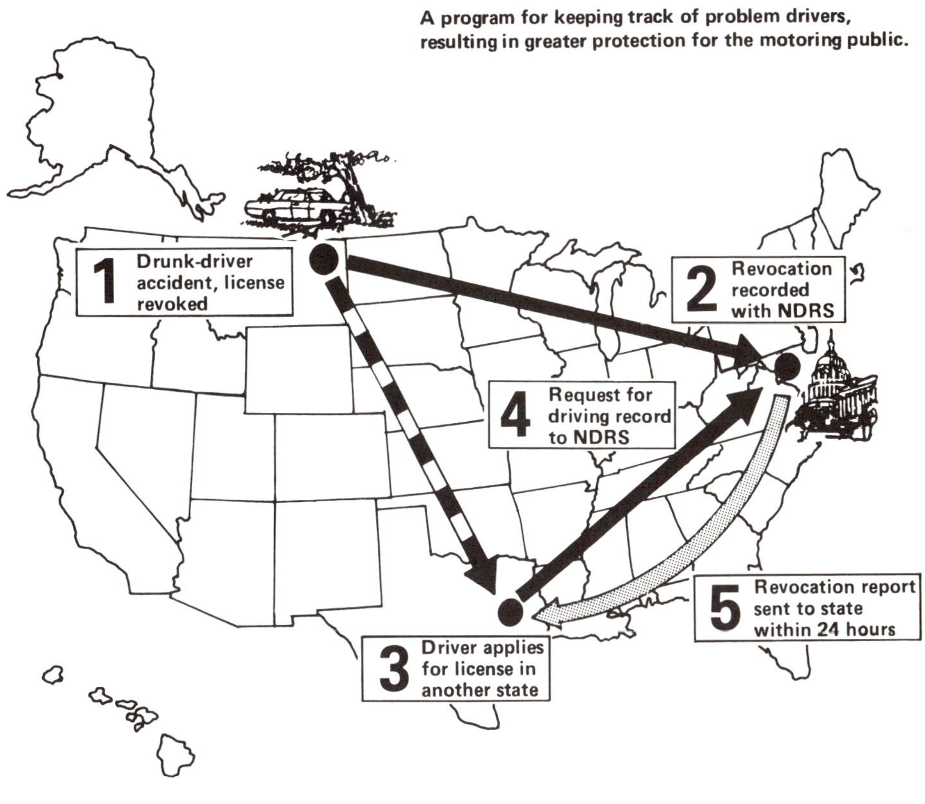

HOW THE NATIONAL DRIVER REGISTER WORKS

2

Getting your car fit for driving

Every car on the road must be registered with the state's department of motor vehicles. The car must be registered so the police or the state knows who the owner is.

All cars on the road must have license plates. In some states, they must have license plates only on the rear. Other states require plates on the front and rear.

The license plate number identifies the car. Every car has a different number. The department of motor vehicles records the number. By looking up the number, the department can learn the name of the owner. It can also find out key details about the car.

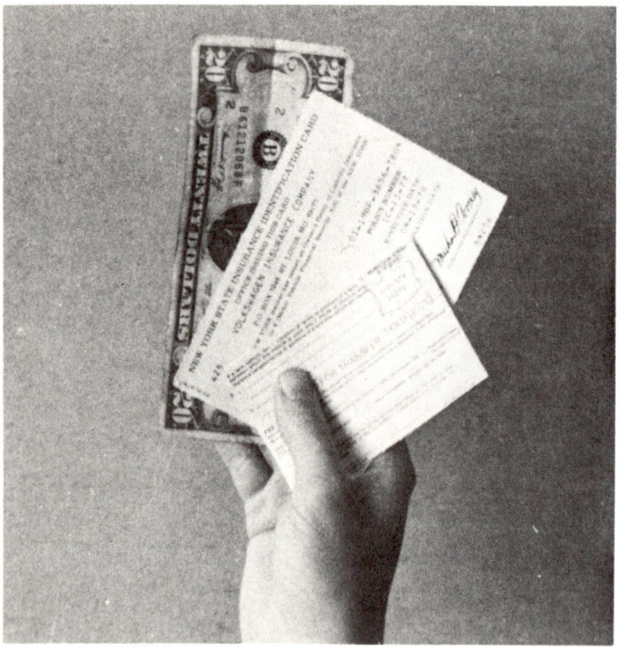

This is a winning hand when you're registering the car.

Getting it registered

You must pay a fee to register a car. The fee varies from state to state. In some states, the rate is based on the weight of the car. In some states, there is a flat fee for passenger cars.

Before you register the car, you may have to prove that you have car insurance. Your insurance company will give you a card to prove you have insurance.

You will also have to prove that you own the car. If you just bought the car, you will have to show the title. This is a paper signed by the former owner. It says you are now the owner.

Every year you must re-register the car. Once you have registered the car, you don't need to show the title. You only need to show last year's registration card to prove that you own the car.

You will probably also need to have your car inspected. It must be inspected to be sure that it is safe to drive. The inspection is meant to keep unsafe cars off the roads.

A car should be inspected at an "official inspection station." Some states require that cars be inspected every six months. Other states say cars should be inspected once a year.

If a car fails the inspection, the owner must have the car fixed and inspected again. If the car is not fixed, the car cannot be driven on any public road. It is not legal to drive a car that is not registered or inspected.

Some states do not inspect many parts of the car. The inspectors check the lights, turn signals, horn, and brakes. Other states also check the tread on the tires and the way the wheels line up. They check the exhaust system and other items.

Car inspections may seem like a bother. But they are important to safe driving. Many accidents are caused by mechanical problems. Brakes fail. Steering fails. Signal lights or brake lights don't work. Inspections are supposed to catch these problems.

Equipment requirements

The federal government has laws that car makers must follow. Standards require cars to have the following features:

- windshields made of safety glass
- defroster for the windshield
- windshield washers
- two-speed windshield wipers
- rearview mirror outside on the driver's side
- rearview mirror inside on the windshield
- uniform pattern for shifting automatic transmission
- seat belts for the rear seats
- seat belts and shoulder belts in the front seats
- parking and regular brakes.

Cars must also have taillights, parking lights, turn signals, back-up lights, and headlights. Brake lights, horn, and a muffler are also required. 1986 model cars and later must have a rear center brake light.

More safety

The government, consumer groups, car makers, and insurance companies can make driving safer. Research on safer cars and equipment is being done. State and federal governments set standards and laws for safety.

Many experts say that cars could be much safer. Sometimes research goes very slowly. Often, car makers are slow to make changes. Many experts feel that safety laws are not tough enough. Experts say cars crumple too easily in crashes. The law requires bumpers to withstand crashes at 2½ miles per hour. That is very slow. Few crashes happen at that low speed. Some cars have bumpers that hold up to crashes at 5 miles per hour. Experts say bumpers should hold up to crashes at 10 or 15 miles per hour. Many are fighting for tougher laws.

The government sets standards that car makers must meet. By 1989, all new cars must have automatic safety belts or air bags. Cars made before 1989 are not required to have these safety restraints.

The air bag inflates inside the car during a crash. It protects the driver from hitting the dashboard, steering wheel, or windshield. Air bags can be installed on the passenger side, too. Only the front of the car has air bags.

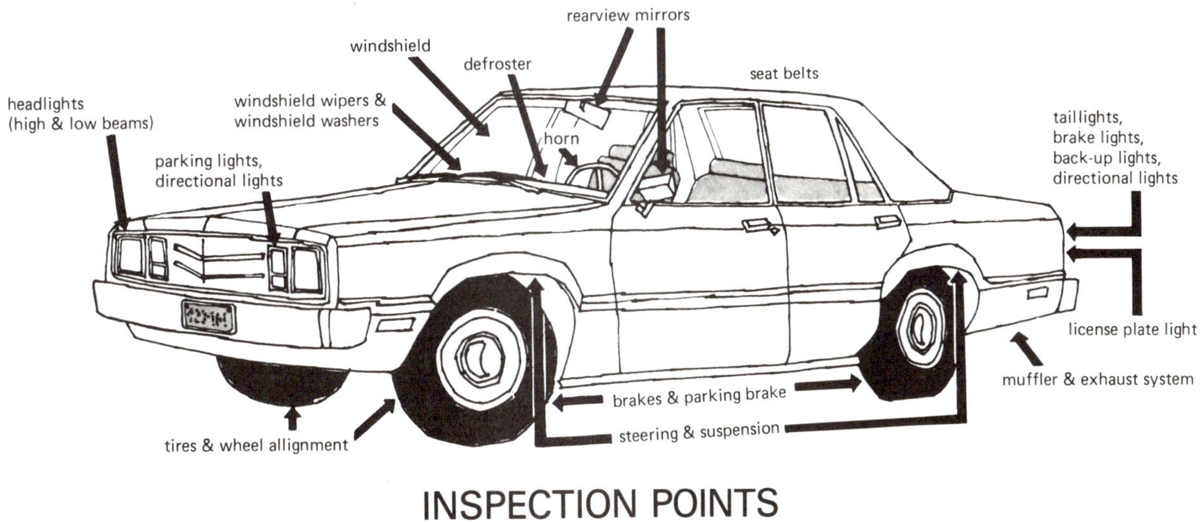

INSPECTION POINTS

The automatic safety belt wraps around the rider when the door is closed. The rider does not have to connect the belt. It works automatically.

States pass safety laws that drivers must follow. Many states require people to wear safety belts. All 50 states now have child safety seat laws. These laws vary from state to state. In every state, the driver must make sure children are safe in the car. Know what the laws are in your state.

Child safety seats save many lives each year. A safety seat protects a child from being thrown from the car during a collision. It also prevents the child from being thrown hard against the inside of the car. A child safety seat can prevent serious injury and even death.

There are many kinds of child safety seats. Some child safety seats are safe. Others do not protect the child well. Government and consumer groups have rated many child seats for safety. Check consumer magazines for these ratings.

Make sure to use a child seat made after 1981. Safety seats made before 1981 may still be sold. These seats may not meet safety standards.

Sometimes, accidents happen because a car is not well-built. There may have been a defect in the brakes or steering, for example. In some cars, the fuel tank is not a safe place. It is not well-protected. It could explode or catch on fire in a crash.

There are other problems in safety design. Often, car makers don't think of simple problems. For example, the designer may put a pointed edge on the trunk door. A person could poke his eye on it.

Cars today are much safer than they once were. But they are still not safe enough. Some consumer groups say that too much money is spent on style. The car makers don't spend enough money on safety research. They have much to do to improve car safety.

Recalls

If there is a safety defect in a certain car model, the car maker must recall those models. The maker must send letters to the owners. The maker must warn the owners and ask them to have a nearby dealer check their cars.

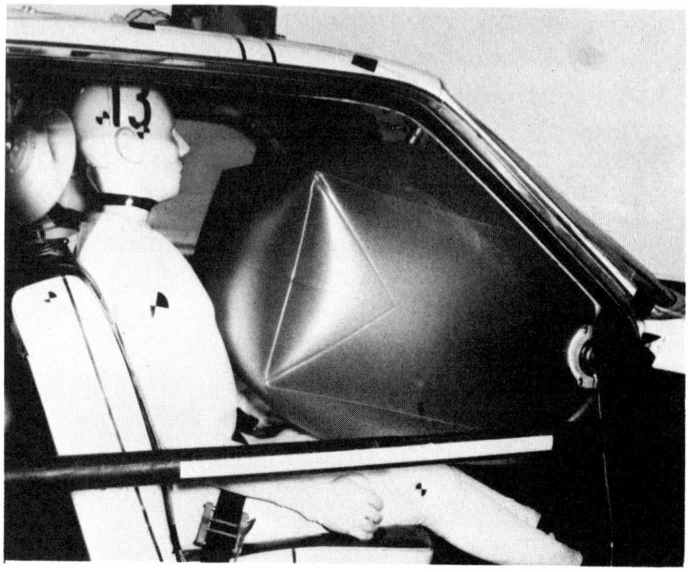

Car makers can't say they've done a bang up job on car safety. But they are improving car safety.

Only a small number of the recalled cars may have the defect. But all recalled cars should be checked. If a defect is found, the maker will pay for the repairs.

There may be a recall of cars that are many years old. The car maker writes the first owner of the car. But that owner may have sold the car. The second owner probably won't get any recall letter. It is too hard to trace the owners of second- or third-hand cars.

Major recalls are announced in the papers and on television. But what if you didn't see the recall announced? And what if you suspect your used car has a recall problem? Check with a dealer or write the company that made your car. You can also call the auto safety hotline (1-800-424-9393). This hot-line gives information about recalls and other auto safety problems.

3 Basic rules you need to know

As you read this book and your state's driver's manual, you will find many rules and tips for driving. There may seem to be a lot to remember.

But almost all the traffic rules are based on only a few basic rules. Let's look at those basic rules.

1. Drive on the right-hand side of the road. Drive to the right of the center line on any street or highway. Stay in the extreme right-hand lane on your side of the road unless:

— you are going to pass another car

— you are going to make a left-hand turn.

You may also drive in a left-hand lane if you are driving on a one-way street.

Always drive to the right of oncoming traffic. Oncoming traffic is traffic coming towards you.

2. Pass to the left of cars going in your direction. We will talk about the few exceptions to this rule later.

3. Signal to tell others what you will do. Don't make other drivers guess what your next move will be. Signal when you plan to change lanes, turn, pass, pull over, or stop.

4. Drive at a safe and legal speed. Drive at or below the speed limit. Never drive faster than the speed limit. Drive slower than the speed limit if traffic or weather is bad. Try to keep pace with the rest of traffic. Driving much slower or faster than the rest of traffic is dangerous for you and others.

5. Allow vehicles trying to pass you to do so. Slow down, if needed, to let them get safely past.

6. Obey right-of-way rules. (See section below.)

Driving fast is dangerous, but so is driving too slowly.

Right-of-way

There are millions of cars on the road. Traffic flows smoothly, usually. Why? How do drivers know who can go first across an intersection? How do drivers know when to stop and when they can go?

Even at intersections with stoplights or stop signs, there are decisions to make. A driver must decide if he can go, or if he must let another car or pedestrian go first. Drivers must follow right-of-way laws if traffic is to flow smoothly.

Right-of-way laws tell which car may go first and which car must yield. *Yield* means to give way to another car or pedestrian.

Learn these rules. They are important to safety and to keeping traffic moving.

1. Suppose you come to an intersection with no stoplights or signs. Another car comes to the intersection at the same time. The car on the right goes through the intersection first.

2. You must yield the right of way if a sign tells you to yield.

3. You must yield to pedestrians at intersections with stop signs or flashing red lights.

4. If the light has turned green, you must yield to pedestrians already in the street.

5. If you are turning left, you must yield to oncoming traffic going straight.

6. When coming out of a driveway or alley or private road, you must yield to traffic on the street.

7. You must yield to emergency vehicles–police cars, ambulances, fire engines–if their sirens and lights are on.

8. You must yield to cars already in a traffic circle if you are just entering the circle.

9. What if you are turning right on a red light? You must yield to pedestrians and traffic moving on a green light.

From city to city

Laws within a state are the same from city to city. If they were different, drivers would be confused. The state has the power to set speed limits in different areas. The state can set the rules which drivers must follow when driving in that state.

(The state may not set speed limits faster than 55 miles per hour. That is the national speed limit. It was set by the federal government.)

From state to state

Laws in most states are pretty much the same. And most are based on the basic rules just described.

There may be some differences from state to state. For example, one state may say you must be within one foot from the curb when parked. Another state may say you must be only six inches from the curb. One state may say you must signal 200 feet before a turn. The signaling distance in another state may be only 100 feet.

So you must learn the laws in your own state. You must study your state's driver's manual.

But for the most part, you can count on the laws being the same from state to state. Having the same rules keeps people from being confused when driving in another state. You still know how to drive smoothly through the streets and highways.

States use a book called the Uniform Traffic Code for making up rules. This code also sets up state guidelines for registering and inspecting cars and for licensing drivers.

Know the signs and signals 4

Traffic signs and drivers' signals help keep traffic flowing smoothly. Signs and signals tell you when to stop, go, and turn. They warn you about possible dangers. They tell you what to expect.

Drivers' signals

Drivers and bike riders must warn each other about their actions. They must signal for turns and for stopping. Drivers can use the signal lights on their cars to signal for turns. Or they may use hand signals. Bike riders must use hand signals. To pass your driver's test, you must know the hand signals.

To signal for a right turn, hold your arm up:

To signal for a left turn, put your arm straight out:

To show you are stopping, bend your arm downward:

Traffic lights

You probably know what the colors of the traffic signals mean. Red means stop. Green means go. Yellow means slow down. Yellow warns you that the red light is about to come on.

At some intersections, there is no yellow light. There are only red and green lights. If both red and green are shining, you are getting a warning. It is the same warning the yellow light gives. It tells you the red light is about to come on.

At some busy intersections, there are extra signals. They are arrows. A green arrow means you may go, but only in the direction in which the arrow points. If you are turning, you do not have to worry about other traffic. A red arrow means stop. A yellow arrow is a warning. But these orders apply only to the lane that the arrow is controlling. And they apply only to cars moving in the direction to which the arrow points.

Some intersections have flashing lights. Flashing lights are used at intersections that are not too busy. A flashing red light means you must come to a full stop before going through the intersection. You may go if there are no cars coming from the right or the left.

A flashing yellow light means you should slow down. You do not have to stop. But you should be ready to stop if necessary. Look both ways before crossing the intersection.

Flashing yellow arrows mean cars going in the direction of the arrow should slow down. Go carefully. Flashing red arrows mean cars going in the direction of the arrow should stop before the intersection. These cars can go through the intersection when it's safe to do so.

Lane control signals

In some places, there are lane control signals. They control the flow of traffic on lanes that change direction at certain times. Some lanes allow traffic to move in one direction at certain hours and the opposite direction at other hours. Signs and signals over the lanes give you orders. A red X signal means do not drive in the lane under the signal. A yellow X means clear the lane. The direction of traffic allowed will soon change. A green arrow pointing down to the lane means you may drive in that lane.

Signs

Road signs give warnings, directions, and information. Each kind of sign has a different shape. Each kind of sign has a different color or colors. You must know the colors and shapes to pass the written part of your driver's test.

Stop sign. A stop sign has eight sides. It has white letters on a red background. A stop sign tells you to come to a full stop before going through the intersection. Make sure the way is clear before you go.

Warning signs. Warning signs are yellow and black. Most are diamond-shaped. Many warning signs use pictures or symbols instead of words. A five-sided sign warns you about a school crossing. A round sign warns about a railroad crossing. Some yellow or orange signs suggest speed limits. These limits aren't law, but obey them. You could be arrested for going faster than conditions allow.

Law signs. Law signs give orders. These are also called regulatory signs. They tell you what you can and cannot do. Most are white and black. Most are rectangles. New ones have symbols, and words below the symbol. These signs give orders about speed, parking, turning, and so on.

A yield sign is a law sign. But it is a triangle. Newer yield signs are red and white. Old ones are yellow and black. Yield means you do not have to come to a full stop if the road is clear. But if the road is not clear, you must stop and wait your turn.

A Do-Not-Enter sign is red and white. It keeps traffic from entering a one-way street or expressway ramp from the wrong direction.

Information signs. Information signs are blue, green, orange, or brown. They tell you about service areas or restaurants. They tell you about distances, exits, and entrances on the freeway. They give route or highway numbers. They tell you about construction projects on the road.

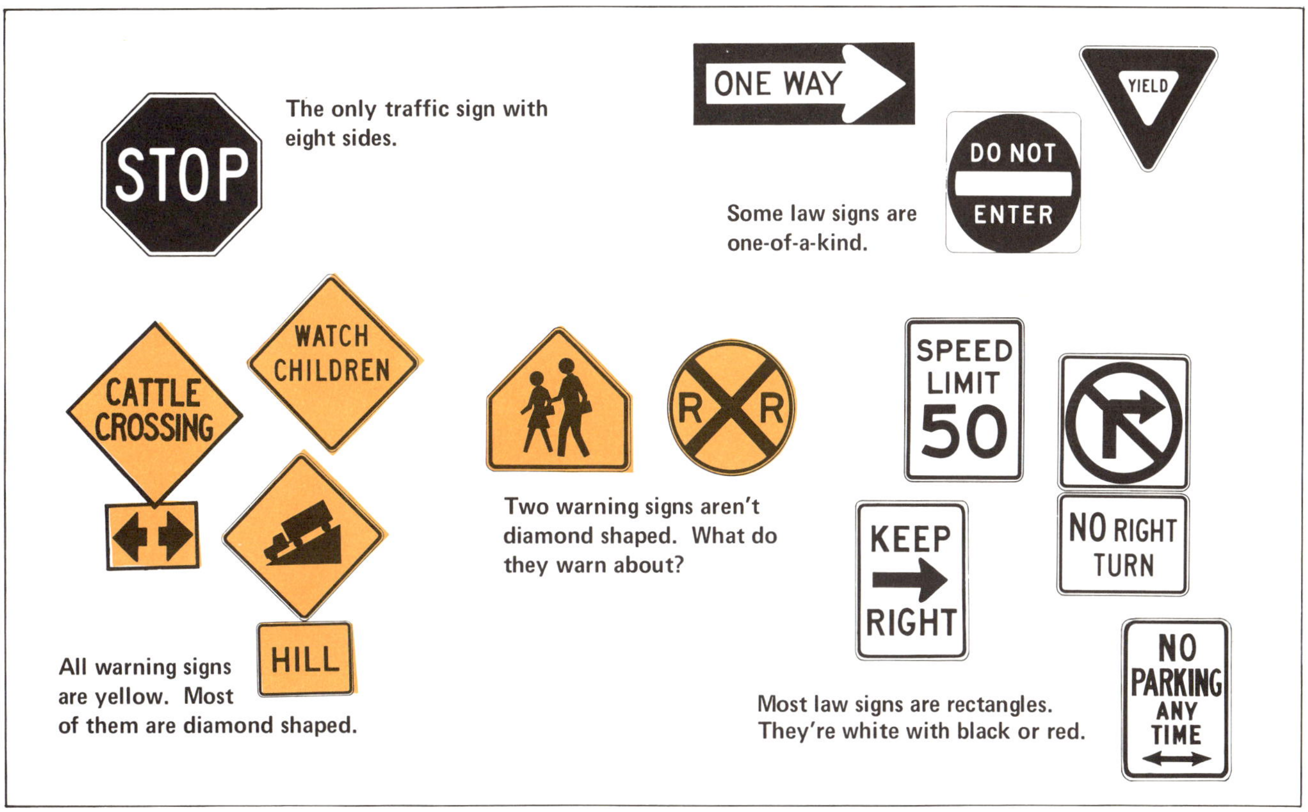

The only traffic sign with eight sides.

Some law signs are one-of-a-kind.

Two warning signs aren't diamond shaped. What do they warn about?

All warning signs are yellow. Most of them are diamond shaped.

Most law signs are rectangles. They're white with black or red.

Reading between the lines

The road pavement is marked with lines. The line tells you what you can or cannot do.

A yellow line separates lanes of traffic moving in opposite directions. A white line separates lanes of traffic moving in the same direction.

A broken white line means you may change lanes if you do not interfere with traffic.

A solid white line means you should stay in your lane. But you may cross the line if it is safe.

A broken yellow line means you may pass if it is safe to do so. (You will find this kind of line on a two-lane highway.)

A double solid yellow line means cars from either direction cannot pass. Only cars making left turns can cross these lines.

What if there is a solid yellow line and a broken yellow line? You may not pass if the solid yellow line is on your side of the broken line. But you may cross the lines to make a left turn. If the solid yellow line is on the other side of the broken line, oncoming traffic may not pass. But you may pass if you will not interfere with oncoming traffic.

Crosswalks are painted across the lanes at many intersections. You must stop before a crosswalk when signs or signals tell you to stop.

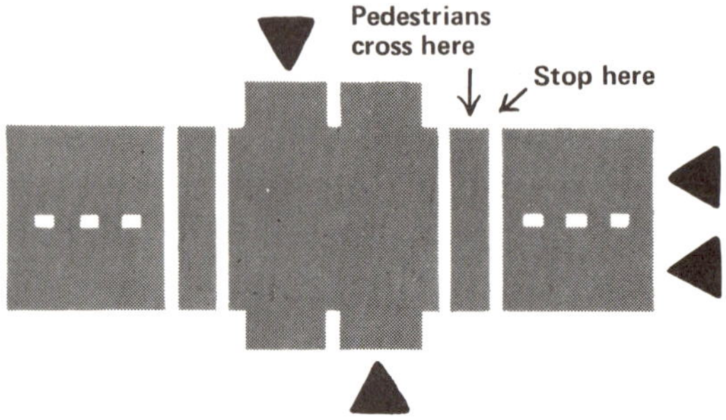

Stop lines are white lines across the lane. They are at many corners with stop signs and traffic signals. Cars should stop before they reach the stop line.

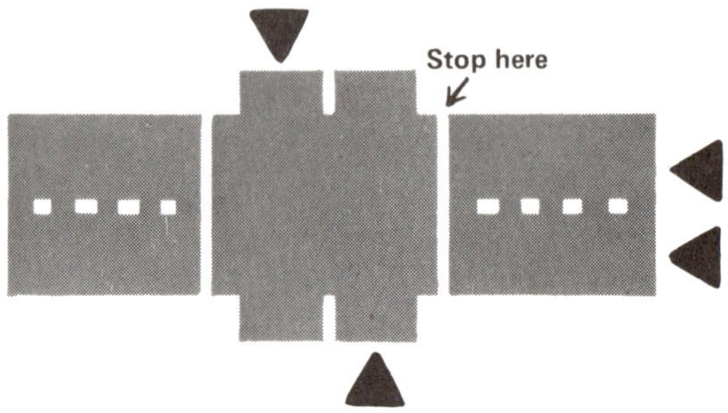

Arrows are sometimes painted in the lane. They show which lane to use to make a right or left turn or to go straight.

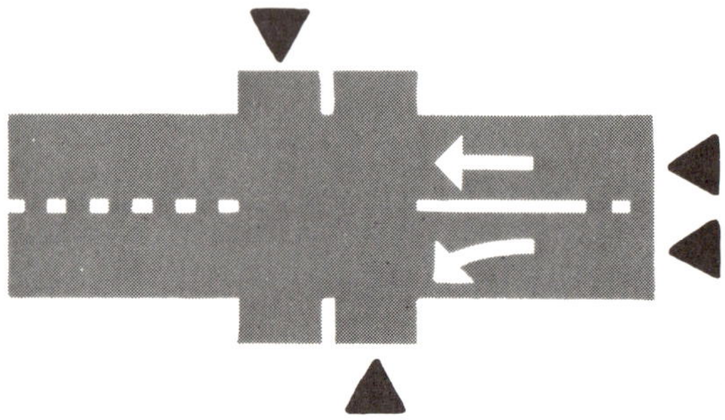

Solid white lines on the side of the road mark the edge of the pavement. If the line angles inward, the road is narrowing. There is a change in the number of lanes. Be ready to merge with traffic from the other lane.

5

Get used to the controls and gauges

Before you drive any car, learn where all the gauges are. Learn how to use the controls. Find out how to read each gauge. Get a good feel for the car you are driving.

Gauges

Speedometer. The speedometer shows how fast your car is moving. Speedometers on many new cars tell both miles per hour (mph) and kilometers per hour.

The speedometer helps you stay within the speed limit. You don't have to guess how fast you are driving. But many speedometers are not accurate. They may be as much as five miles per hour wrong. They are usually wrong on the slow side. That is, when you are driving at 60 mph, the speedometer says 55 mph. So you don't even know that you are speeding.

If you find that you are passing a lot of other cars, you are probably driving too fast. If your speedometer says you are driving at the speed limit, have the speedometer checked. Or maybe the speedometer says you are driving at the legal speed, but everyone is passing you. Maybe they are speeding. But maybe your speedometer is wrong. Have it checked.

Tachometer. Some cars have tachometers that tell how fast the engine is running. This helps you know when to shift gears. You shift when the engine is running too fast or too slowly.

Odometer. The odometer is on the face of the speedometer. The odometer counts off the miles the car travels. It shows the total number of miles the car has driven.

You can use the odometer to figure out many things:

- how far you have driven on a trip
- gas mileage
- if it is time for maintenance work
- how far you are from the next road, exit, or town.

Gas gauge. The gas gauge tells you how much gas is left in the tank. Keep an eye on this gauge. Try not to let the tank go below one-quarter full.

Temperature gauge. The temperature gauge shows how hot or cold the engine is. There is a hot zone, usually colored red. And there's a cold zone, often colored blue. The area in the middle is the normal or safe zone.

The needle stays in the cold area until the car is warmed up. It slowly moves to the middle zone. If the needle moves to the red zone, the engine is too hot. Don't drive the car when it's this hot. Stop and let it cool. Then take the car to a mechanic. Find out why the engine is too hot.

Some cars have only warning lights to tell you when the car is too hot.

Ammeter or warning light. The ammeter is a gauge. It shows how much electricity is going to the battery, or being taken away from the battery. The gauge has two letters. D means discharge. C means charge.

The needle on the ammeter points to D when you start the car. But if the needle stays at D when the engine is running, there is a problem.

Many cars do not have an ammeter. They just have a warning light. The light goes on when you start the car. But it should go off right away. If

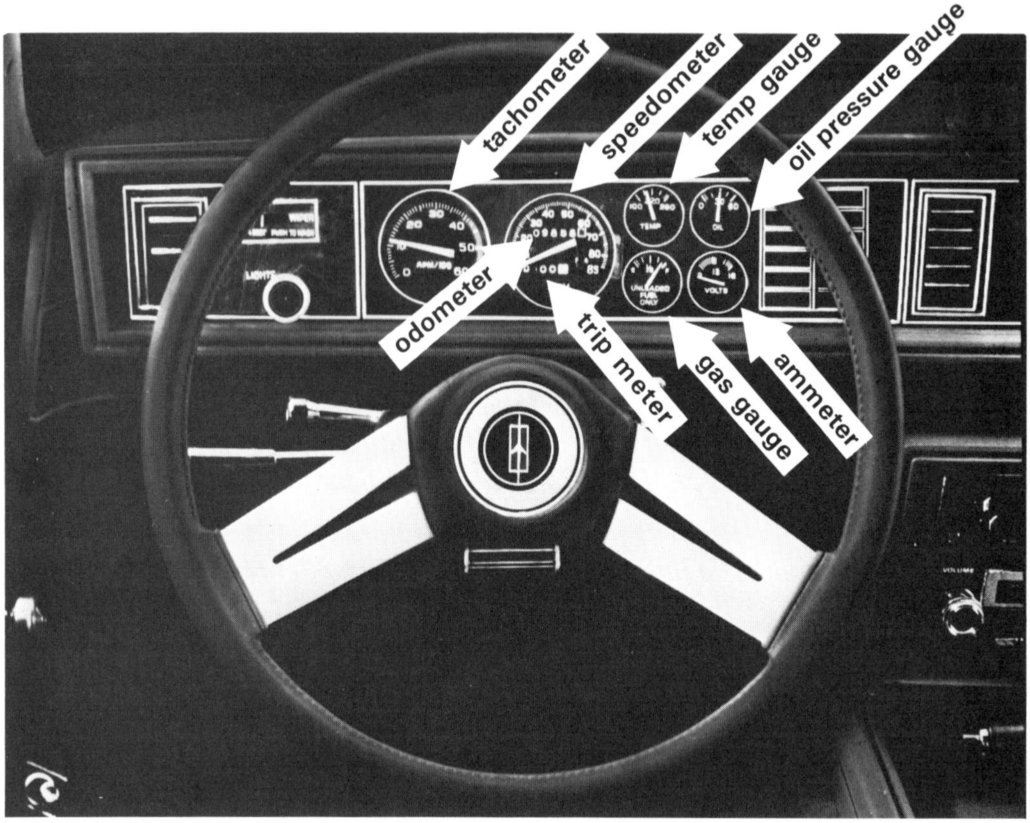

it stays on while you drive, find out the reason. The light means electricity is being drained from the battery.

Oil pressure gauge or light. Some cars have an oil pressure gauge. The gauge shows whether the oil is being pumped through the engine at the right pressure. The gauge gives you early warning if the oil pressure is too low. It warns you before the problem becomes an emergency.

An oil pressure warning light is also called an "idiot light." The light doesn't give you much advance warning about a problem. If the light comes on when you are driving, pull over right away. Your oil pressure problem is probably already serious. Turn off the engine. If you don't, you could ruin the engine. Some parts of the engine are not being oiled.

The oil pressure light flashes when you first start the car. But it should go off right away.

Lights

Headlights. You turn on the headlights by pulling a switch on the dashboard. After the headlights are on, you can turn them from low to high beam. High beam helps you see farther ahead than low beam. Different cars have different ways to change beams. On some cars, you use the lever that controls the turn signals. On other cars, you push a switch on the floor.

Headlights must be used between sunset and sunrise. They also should be used in poor weather—fog, rain, and snow.

High beams can be used at night when there is no one near you. But you must change to low beam for cars in front of you and cars coming toward you.

High beam indicator light. A red or blue light glows on the dashboard when the high beams are on.

Parking lights. If you pull your headlight switch part way out, the parking lights turn on. Use them when parked in places along the road where cars don't usually park. This warns other drivers to avoid your car.

Turn signals. You can warn other drivers you are going to turn by using the turn signals. They are worked by a lever on the left side of the steering column. To signal a right turn, push the lever up. For a left turn, push the lever down.

This lever makes lights on the outside of the car blink until the turn is made. A green light on the dashboard also blinks to tell you the turn signal is on.

If you are only changing lanes, or if the turn was not sharp, you may have to turn off the signal yourself. Simply move the lever to the middle position.

Brake lights. Red brake lights go on when you push the brake pedal. The lights are at the rear. The lights warn other persons that you are slowing down or stopping.

Back-up lights. White lights go on at the rear when the car is in reverse. They tell others you are backing up. They also help you see at night when you back up.

Four-way flashers. When the four-way flashers are on, both the right and left turn signals in the front and rear blink. They warn other persons that the car may be a hazard to them. The car may be stopped by the side of the road. Or the car may be moving very slowly. On most cars, the control button for the flashers is on the steering column. Turn off the four-way flashers when signalling to turn. If you don't, other drivers won't know which way you're going.

Driving controls

The operating controls are the ignition, steering wheel, brake pedal, gas pedal, clutch, gearshift lever, and parking brake.

Ignition. The ignition usually has three positions. One position locks the steering wheel. A second position starts the car. As you turn the key, you press on the gas pedal. The engine starts running and you let go of the key. A third position allows you to play the radio or use the heater fan when the engine isn't running.

Steering wheel. The steering wheel is easy enough to find! It's right in front of you when you sit in the driver's seat.

You control the steering wheel. The steering wheel controls where the car goes.

Some steering wheels can be adjusted. The driver can move the wheel up and down. She can move it to the position that is most comfortable.

Brake pedal. The brake pedal is near the floor. On a car with an automatic transmission, the brake pedal is the pedal on the left. On a standard transmission car, the brake pedal is the middle pedal. You use the brakes to slow down and stop the car. You use your right foot to work the brake.

Gas pedal. The gas pedal is the floor pedal on the right. It is worked with the right foot.

When your foot is on the gas pedal, your heel rests on the floor. To go faster, you push down on the pedal with the ball of your foot.

Clutch pedal. Cars with a standard transmission have a clutch pedal. This is the pedal on the far left. It is worked with the left foot. You push down on the pedal when you change gears.

Gearshift control. The stick or lever for changing gears may be on the right side of the steering

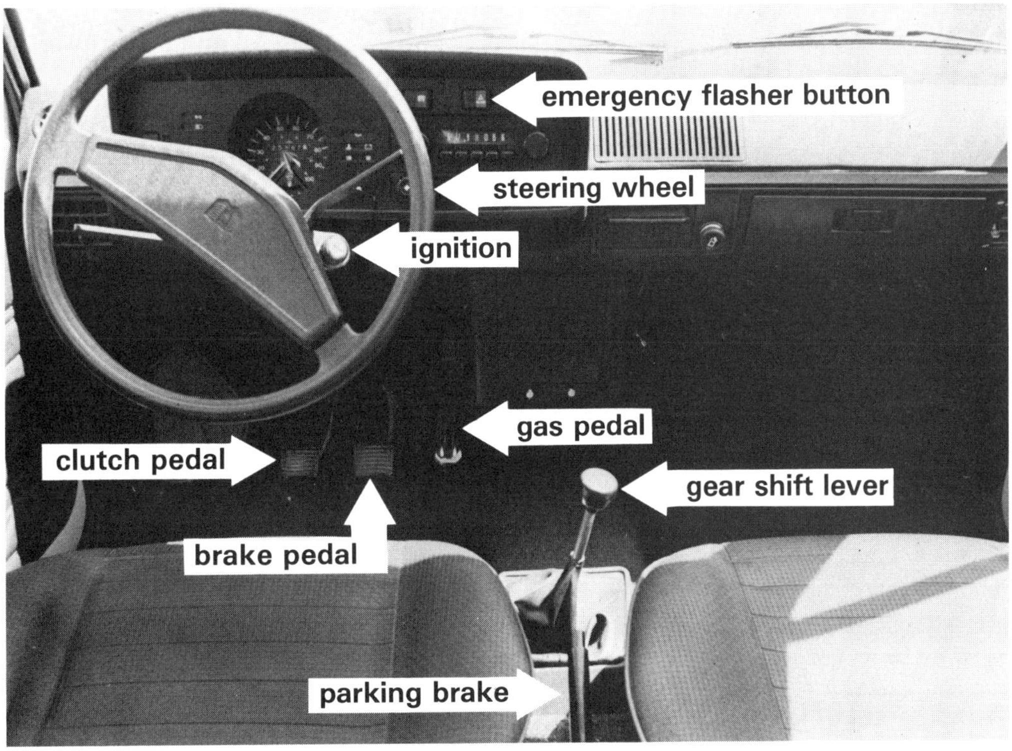

column. Or it may be on the floor between the two front seats. It is worked with the right hand.

On cars with a standard transmission, you must push in the clutch when you move the gearshift lever. Standard transmissions have three, four, or five forward speeds. There are also reverse and neutral positions.

To change gears on a car with automatic transmission, just move the lever. There are five or six positions for the lever. They are park, reverse, neutral, drive, and low. (There may be one or two low speeds.)

Parking brake. The parking brake is also called an emergency brake because you can use it if the brakes fail. It is usually used with the parking gear to keep the car from rolling when it's parked.

On some cars, the parking brake is controlled by a lever between the two front seats. On other cars, it is worked by pushing a foot pedal on the far left. It is released by a handle near the foot pedal.

Comfort and safety controls

Windshield wipers. In most cars, the switch for the wipers is on the dashboard. In some cars, it is on the turn signal lever. The windshield wipers have two speeds.

There is also a windshield washer unit. The switch for the wipers controls the washer. (On older cars, the washer switch may be on the floor.) When you work the switch, water and cleaner is sprayed onto the windshield.

Horn. On most cars, the horn is on the steering wheel. The horn should be used only to signal or warn other drivers or pedestrians.

Heater/defroster. The heater blows warm air into the car. The defroster blows warm air onto the windshield. The defroster keeps the windshield clear of ice and frost.

The heater and defroster are worked by levers on the dashboard. Some cars also have a rear-window defroster to keep the rear window clear. It is worked by a switch on the dashboard.

Rearview mirrors. There is a rearview mirror inside the car. There is one outside on the driver's side. There may be one outside on the right side, too. The mirrors help you see behind you and behind you to the side.

Some inside mirrors have day and night settings. The night setting cuts down the glare from the headlights of cars behind you.

Sun visors. The sun visors are over the windshield. You can pull the visors down. Some also swing to the side. They keep the sun out of your eyes. When you adjust the visors, make sure the visors don't block your view of the road.

Radio. Many cars have radios. A car radio has push buttons for quick tuning. You can set the buttons to your favorite stations. To set the buttons:

— tune in to the station you want
— pull the button all the way out
— push the button in again.

Now, each time you push that button, you will tune into the same station.

Practice

While you are sitting behind the wheel, practice a few important moves.

Practice moving your foot from the gas pedal to the brake pedal. Get a feel for the brakes. Do you have to push far down to feel any tension? (Before you drive a car, find out if it has power brakes. If so, a gentle touch will slow down the car.)

Practice reaching for different controls—light switches, wiper switch, turn signals, and so on.

Practice glancing at the speedometer. You should check this gauge often when you drive. Look at it quickly. Then look back to the road.

Practice looking in the rearview mirrors. When you are driving, look in the mirrors often, about every 20 seconds. Take a quick look. Then look to the front again. Never take your eyes from the road for more than an instant.

6 Count down! Last minute details

Before you drive anywhere, take a moment to make yourself comfortable and safe behind the wheel. Take these steps:

1. Check the seat position. Adjust the seat position if you need to. The seat moves backward and forward. Some seats move up and down. The back of the seat can be adjusted, too.

There are many seat positions. Choose a position that lets you reach the controls easily. You shouldn't have to stretch. You shouldn't be too cramped either.

When you drive on the highway, you may want to sit farther back than you do for city driving. Then you can stretch your legs out more. You will be more comfortable for a long ride.

Why buckle up?

2. Fasten safety belts. Every car must have safety belts. Take the time to fasten them. The shoulder and seat belts can save you from serious injury and even death.

Even in a minor collision or a quick stop, safety belts can save you from injury. They keep you behind the wheel where you belong so you can control the car.

In a serious collision, safety belts can keep you from hitting the windshield. They keep you from being thrown from the car. Even if your car has air bags, safety belts give you added protection.

3. Fasten children in safety seats. Safety seats protect against injury and death. During a collision, a child can be thrown out of the car. The crash could throw the child very hard against the inside of the car. The child could die or be injured very seriously. Even a low-speed crash could be fatal for a child. Almost 80% of child deaths in cars could be avoided by using safety seats.

Always put a child in a safety seat. And use the safety seat correctly. When used incorrectly, the seat may put the child in danger.

If there is no child safety seat, use the safety belt. Use the belt only if the child can sit up by herself. Make sure the safety belt fits well. The belt should be as low on the hips as possible. Don't use the shoulder strap if it crosses the child's face or neck. Put the shoulder strap behind the child's back after it is fastened.

4. Lock the doors. If the doors are locked, there is less chance they will spring open in a collision.

A crash with no belt | with lap belt | with lap and shoulder belt

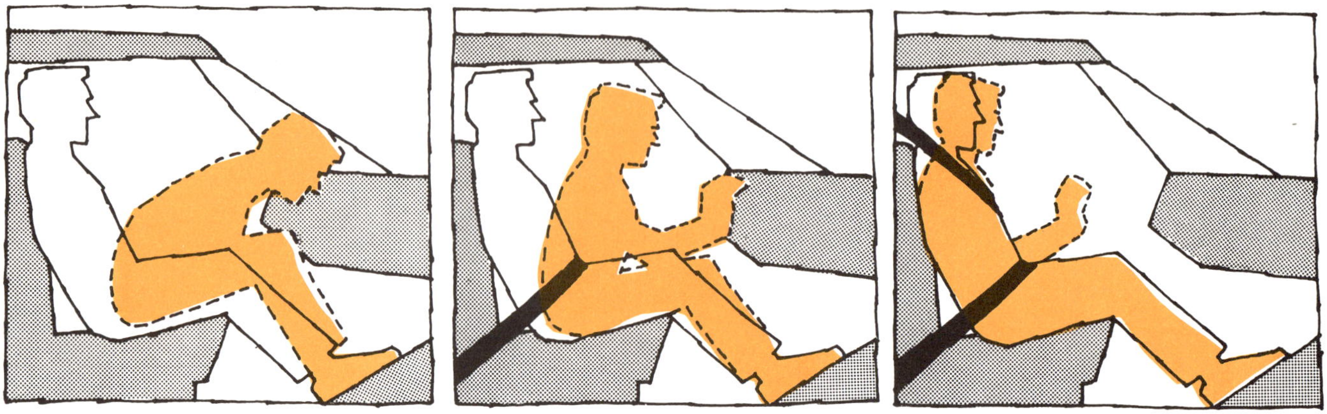

So there is less chance of your being thrown from the car.

Locked doors also keep people from entering your car at a red light or stop sign.

If doors are locked, children will not open them easily.

5. Adjust the mirrors. A blind spot is an area that is blocked from the driver's vision. It is blocked by part of the car. Blind spots are usually to the rear or the side. The driver cannot see objects or persons in the blind spots. But using rearview mirrors gets rid of some blind spots.

The rearview mirror inside the car should be adjusted so you can see directly behind you.

The outside mirror on the driver's side should be adjusted, too. Adjust it to help you see cars moving up behind you in the left lane.

Some cars have an outside mirror on the passenger side. That mirror should be adjusted to help you see cars moving up in the right-hand lane. This mirror is very helpful when you must merge with traffic on your right.

Mirrors help you to see. They shouldn't get in your way. Make sure the mirror doesn't block your view to the front. A mirror shouldn't be at your eye level. If it is, it will block out a spot to the front or side.

6. Adjust the heat, vents, or air conditioning before you get the car moving. Don't turn the heater fan or air conditioner on before you start the car. This would make the battery work too hard. But adjust the levers that control how hot or cold you want the car to be. Then you won't have to bother so much with these levers while you're driving.

The view with inside mirror | with outside mirror | with both mirrors

Part Two

You know the basic rules of the road. Your car's in shape. You've got an idea of what to do with the car. Now you need skills. So let's talk about **the important skills** you'll be using when . . .

You're on the road!

Using an automatic transmission 7

If your car has an automatic transmission, you don't have to work very hard to shift gears. You simply have to move the gearshift lever to one of the five or six positions:

P (park). The car can be started with the lever at P. The lever should be in this position when the car is parked. Then the car can't roll.

R (reverse). Reverse is used for backing up the car. The car can't be started with the lever in this position.

N (neutral). The engine is not connected with the wheels when the lever is in this position. So, if the engine's running, it won't move the car. The engine can be started when the car's in neutral. And the car can roll if it's on a hill or being pushed.

D (drive). The car moves forward with the lever in this position. Most driving is done with the lever at D. The car cannot be started with the transmission in drive.

L (low). The low gear or gears give extra power at slow speeds. Some cars have two low speeds. They may be called first and second, or L_1 and L_2. The low gear or gears help the car move up a steep hill. Low gears also slow down the car. This is important for going down steep hills. First gear, or L_1, gives even more power at slow speeds than second gear, or L_2. It would also slow down the car more when you go down a hill.

The gears follow the same pattern in every automatic transmission car. You can see the gear positions in the photos on this page.

On some cars, the gearshift lever comes out from the steering column. It's the lever on the right.

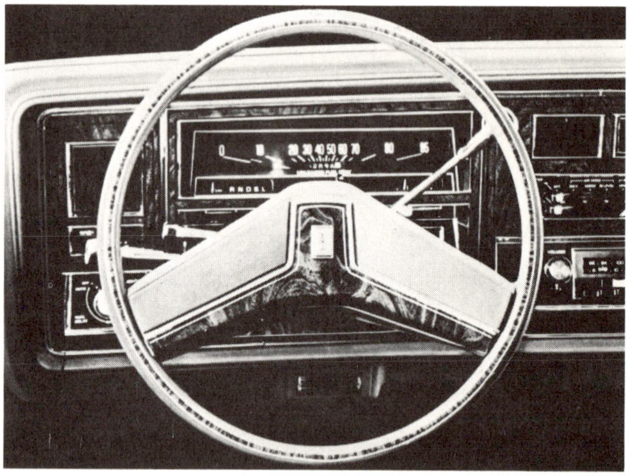

On some cars, the gearshift lever is between the seats.

Starting the car

To start a car with an automatic transmission:

1. Be sure the parking brake is on.
2. Make sure the gearshift lever is on P.
3. If the engine is cold, push the gas pedal to the floor once, or twice for some cars.

4. Put the key in the ignition. As you press gently on the gas pedal, turn the key all the way to the right. As soon as the engine turns over, let go of the key. (On some cars, you don't push on the gas pedal when you turn the key.)

Let the engine warm up for a minute or two.

Read the car owner's manual if you have one. Find out what the car maker says about starting the car. Some cars have slightly different starting steps. There may be special steps for starting a cold engine. There may be special steps for starting a warm engine.

Moving forward

How do you get a car with an automatic transmission to move forward?

1. Press down on the brake pedal with your right foot.
2. Put the gearshift lever on D.
3. Release the parking brake.
4. Look around to check traffic. Use mirrors. Then look over your shoulder to check behind and to the sides.
5. Signal which way you intend to move.
6. Take your foot from the brake and put it gently on the gas pedal. To increase your speed, push down harder on the gas pedal. The car may start to inch forward before you put your foot on the gas pedal. So don't take your foot off the brake until you are ready to move.

Backing up

To back up:

1. Push down on the brake pedal.

Check behind you before moving out onto the street.

When backing up, sit so you can see well out the rear window. But make sure you can handle the controls.

2. Put the gearshift lever on R.
3. Release the parking brake.
4. Check traffic.
5. Turn around to check behind and to the sides. Look over your right shoulder to watch where you'll be going. Steer carefully.
6. Take your foot off the brake when you are ready. The car will probably start to move. If it doesn't, push gently on the gas pedal. Move slowly. Be ready to stop if you have to.

Stopping the car

Before stopping the car, check the traffic behind you. Look in your rearview mirror. Make sure there is no one too close behind you.

If it is safe to stop, touch your brake pedal lightly once or twice. This flashes on your brake lights.

Take your foot off the gas pedal. Put your foot on the brake pedal. Slow down gradually. Just before the car stops, ease up on the brake pedal a little. The car will come to a smoother stop.

Start braking soon enough before you want to stop. Then you won't have to make a quick stop. Judging when to start braking takes practice. It depends on how fast you are moving, and on how good the brakes and tires are. It also depends on the condition of the road.

Braking with the left foot

Some drivers use their left foot to brake when they are parking or holding the car on a hill.

Parking in a tight spot. Left-footed braking is useful when you try to park in a tight place. You do it when the car is inching forward or backward.

Put your left foot on the brake pedal. Work the gas pedal with your right foot. Put light pressure on the gas pedal. Little by little, let up on the brake pedal. Let the car move forward, or backward, very slowly and steadily. Keep light pressure on the gas pedal when you release or push on the brake pedal. Release the brake pedal, push on the brake pedal, release, and so on. This lets the car inch forward or backward.

Holding the car on a hill. After stopping on a hill, move your left foot to the brake pedal. Don't take your right foot from the brake until your left foot is holding down the brake. Then put your right foot over the gas pedal.

When it is time to move further up the hill, push on the gas pedal. Then take your foot off the brake pedal. The car will move forward. (Don't push too hard on the gas pedal or the car will leap forward.)

Starting on a hill

On slight hills, put your right foot on the brake to stop the car. When you are ready to go, move your foot quickly to the gas pedal. This is no different from any other stop-and-go situation.

But, on steeper hills, other methods may be needed. We already talked about the two-footed method. Here is another method.

Come to a stop by using the foot brake. Then set the parking brake. Leave the transmission in drive. Move your right foot to the gas pedal. When you are ready to move, push on the gas pedal. You can feel the engine speed up. Check traffic. Release the parking brake. Push down a little more on the gas pedal to keep the car moving up the hill.

Covering the brake pedal

Some experts say drivers should learn to cover the brake with the left foot. This means the driver should drive at certain times with his left foot over the brake pedal. The driver should be ready to hit the brake at any moment. (The foot should not touch the pedal. Only a light touch will begin stopping a car with power brakes. And even without power brakes, always touching the brake could cause brake damage.)

The experts say the brake pedal should be covered in risky situations. Those situations are:

– heavy traffic
– intersections
– when children are in sight.

What if you don't ever brake with your left foot? Then you should get used to covering the brake with your right foot in risky situations. This means you'll be taking your right foot off the gas pedal. Poise it over the brake pedal. This will slow the car down. You may not need to hit the brake at all. But your foot is ready, just in case.

Low gear

Most of your driving will be done with the gear in drive. But to shift to low, slow down to the speed suggested in the owner's manual. That speed is usually 20 or 25 mph, or slower. After you have slowed down, shift to low.

Use low gear for:

1. Climbing very steep hills.
2. Going down long, steep hills. The low gear slows down the car. The engine holds the car back. Use low, or first gear, for very steep hills. Use L_2, or second gear, for less steep hills.
3. Getting more power at slow speeds. It sometimes helps to use low gear for pulling through mud, sand, or snow, or for hauling heavy loads.

Passing gear

Passing gear is not shown on the console or steering column as one of the gears.

The passing gear is for fast pick-up. To put the car into passing gear, push the gas pedal quickly to the floor. That puts the transmission into *passing* gear for extra power.

Passing gear is useful for gaining speed quickly without shifting out of drive. It is very useful for passing on a two-lane highway.

After you have gained speed, let up on the gas pedal. The car will return to the normal driving gear.

Read the owner's manual for your car. Find out at what speeds the passing gear works on your car.

8
Handling a standard transmission

First, let's review the operating controls of a car with standard transmission. (It's also called manual transmission.)

On the floor, there are three pedals. You'll see the clutch pedal, the brake pedal, and the gas pedal. The clutch is at the far left. You use your left foot to work it. The brake pedal is in the center. You work it with your right foot. The gas pedal is at the far right. You work that with your right foot, too. (You never have to have your foot on the brake and gas pedals at the same time.)

On the steering column, or between the two front seats, is the gear shift lever. A car with standard transmission has three, four, or five forward speeds. There is also a reverse position for moving backwards. And there is a neutral position. The car isn't in gear when it is in neutral. No power gets to the wheels. The engine can't move the car then. But the car can still roll when the lever is in neutral.

Shifting gears

To change gears, push in the clutch and move the gearshift stick. Never move the stick until you have put in the clutch.

There is a pattern to the gears. As you can see, the pattern looks something like an H with a few additions.

Neutral is in the middle of the pattern. There is an easy way to tell if the car is in neutral. Just shake the gearshift stick. If the stick wiggles freely, the car is in neutral. If the stick won't move easily, the car is in gear.

In 4- and 5-speed cars, first gear is at the top-left side of the H. To move from neutral to first, push the stick over to the left and then up.

Four-on-the-floor

On the steering column

To move from first to second, move the stick straight down, past neutral.

To move from second to third, push the stick up to neutral and then over to the right and up.

From third to fourth is easy. Move the stick straight down, past neutral.

Reverse is in different places on different cars. On the four-speed shown above, here's how to shift into reverse. Starting at neutral, push the stick as far to the left as you can. Then push the stick down toward the seat.

Practice shifting with your car in a safe place. Use a place like a level driveway where it won't roll. Practice until you can find the positions without having to stop and figure out where they are. Keep the engine off and the parking brake on while you do this. And, of course, push the clutch pedal to the floor when changing gears.

Move through all the gears. Go from first to second to third to fourth. Then try it backwards. Move from fourth to third, third to second. Practice until the movements become natural for you.

Speeds and the right gears

With experience, you will be able to tell when to shift. You won't have to look at the speedometer or tachometer. You will be able to tell when to shift by the sound of the engine.

Suppose the engine sounds as if it is racing or working too hard. This means you are driving too fast for the gear you are in.

If you drive too slowly for the gear you are in, the car will seem to be chugging. It may even buck a little. And it will stall if you don't speed up or shift to a lower gear.

The chart on this page shows at what speeds you can drive in each gear. The chart also shows at what speed you should shift to a higher or lower gear. The range is about the same in most cars. Check your owner's manual for the best range for your car.

As you can see from the chart, the lower the speed, the lower the gear.

First gear is mostly for starting from a dead stop. You don't stay in first very long after you get moving.

Use the lower gears for going up or down steep hills. In the lower gears, the engine slows down the car when going down hills. This saves wear and tear on the brakes. In the lower gears, more power gets to the wheels. This helps in climbing hills.

Most of your driving will be done in third and fourth gears. Many cars have a fifth gear. Fifth gear helps save gas during highway driving.

When shifting, never skip a gear. Move from first to second to third to fourth (to fifth). Move through the gears in proper order.

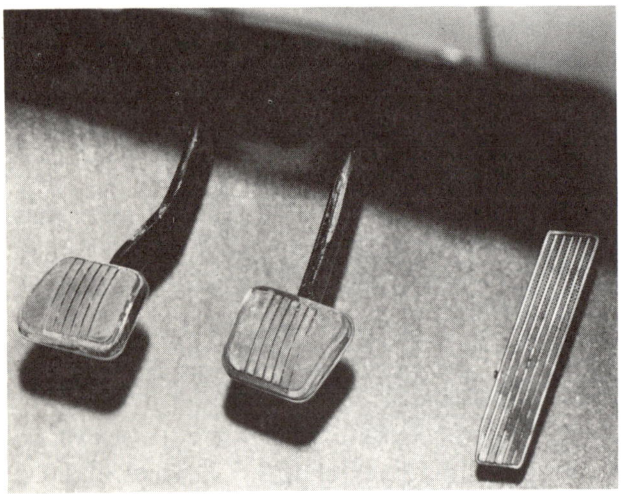

A car with standard transmission has three foot pedals.

3 SPEED		4 SPEED	
GEAR	SPEED	GEAR	SPEED
1	0-10	1	0-10
2	10-20	2	10-20
3	20 +	3	20-45
		4	45 +

Basics to remember before you drive

Here are a few basics to remember before you drive a car with standard transmission:

1. Always push in the clutch before you move the gearshift stick. Push the clutch *all the way to the floor*.

2. Take your foot off the gas pedal as you put the clutch in.

3. Don't rest your foot on the clutch while you are driving.

4. Don't drive too fast or slowly for a given gear. (We will talk about that soon.)

5. Push the clutch in before you come to a complete stop, or the car will stall.

More practice

The movements you will have to know are explained in the next few sections. Practice each of these moves while the engine is off. This will help prepare you for actual driving later on.

Starting the car

To start the car:

1. Be sure the parking brake is set.

2. If the engine is cold, push the gas pedal to the floor once or twice.

3. Put the gear in neutral. Some people simply push the clutch in. When the clutch is in, it's the same as if the car were in neutral. But it's safer to simply shift into neutral. Then you don't need to worry about holding the clutch in while you start the car.

If you start the car when it's in gear, it will lurch forward or backward unless the clutch is in.

4. Put your right foot on the gas pedal. As you press lightly, turn the key to the right all the way. When you hear the engine start, let go of the key and lift your foot off the gas pedal.

If your car has fuel injection, do not press the gas pedal. Just turn the key and let go when the engine starts.

Getting the car moving

To get the car moving:

1. With the left foot, push the clutch pedal all the way to the floor.
2. Push down on the brake pedal with your right foot.
3. Shift into first gear.
4. Release the parking brake.
5. Check the side and rearview mirrors. Check all around you. Make sure the way is clear.
6. Move your right foot to the gas pedal. Push very gently on the gas pedal. At the same time, let out the clutch very slowly. Do this until the pedal reaches the friction point. The friction point is the point at which you can feel the car starting to move. Hold the pedal at the friction point for a moment. Then slowly let the clutch out the rest of the way.

Let the clutch out slowly or the car will stall. Also give just the right amount of gas. Don't use jerky foot movements.

Keeping the car moving

When the car is moving at about 10 miles per hour (mph), it's time to shift into second. Here's how to shift while moving:

1. Put the clutch to the floor.
2. As you push in the clutch, take your foot off the gas pedal.
3. Move the shift stick to second gear.
4. Lift your foot off the clutch.
5. Put your foot on the gas pedal again. This whole shifting movement only takes a second or two.
6. At about 20 mph, shift into third. Follow the steps for shifting into second.
7. At about 40 mph, shift into fourth. Again, follow the steps for shifting into second.

Fifth gear is used mostly on highways. It is easier on your engine at high speeds. And it helps to save gas. Use fourth gear to reach your cruising speed. Then shift from fourth to fifth gear. Cruising means driving at the same speed for a while. Don't use fifth gear unless you are cruising at over 45 miles per hour.

Stopping the car

Here's how to stop the car:

1. Take your foot off the gas pedal. Slow down to a speed where you can shift into the next lowest gear. Always downshift just before you reach a speed that is too slow for the higher gear. Use the chart on page 30 as a guide.
2. Push in the clutch and shift into the next lowest gear.
3. Release the clutch slowly.
4. Repeat steps 1 through 3 until you are in 2nd gear and nearing the stop.
5. To stop, push in the clutch and the brake. Keep the clutch and the brake pedal pushed in until the car stops. Shift into neutral. Let go of the clutch but keep your foot on the brake. To move again, push in the clutch and shift into first gear.

You may only need to stop for a moment at a traffic light or a stop sign. If so, keep the clutch and brake pedals pushed in. But don't bother shifting into neutral. Shift into first gear. When it's safe, you can drive again.

To stop and park, put on the parking brake after the stop. Keep your foot on the clutch. Shift into first gear and turn off the car.

Do not "ride the clutch." This means keeping your foot on the clutch while braking to slow down. Only use the clutch to downshift or when you must stop quickly. Your clutch won't wear out as quickly. When stopping, use your brake as little as possible. Downshifting slows your car without the brakes. Your brakes will have a longer life.

Sometimes you must stop quickly. When you don't have time to downshift, you must use your brake. Push in the brake and the clutch until the car stops.

Shifting on hills and turns

When going down a hill, downshifting helps to control your car's speed.

1. Begin to slow down before you start down a hill. To slow down, downshift as in steps 1-3 above (Stopping the car). Use the brake pedal if needed.

2. After shifting, release the clutch slowly. Press gently on the gas pedal if needed.

When climbing a hill, you may have to downshift to keep your speed. Shift quickly so you don't lose power.

Slow down for turns by downshifting. If you are turning a corner at less than 20 mph, shift into second. Shift before turning. You need both hands to turn the steering wheel.

Backing up

To back up a car with standard transmission:

1. Make sure the parking brake is on.

2. Push down the clutch

3. Put your right foot on the brake pedal. Release the parking brake.

4. Shift into reverse.

5. Check traffic in all directions. Follow the safety rules for backing up as given in chapter 14.

6. Move your right foot to the gas pedal.

7. Put slight pressure on the gas pedal. Start to let the clutch out as far as the friction point.

8. Push down on the gas pedal a little more. Let the clutch out the rest of the way—slowly.

To stop the car, push in the clutch, then put your foot on the brake pedal. If the clutch is not in, the car will stall.

Starting on a hill

Starting your car on a hill is one of the hardest moves in a car with standard transmission. The car rolls back when you are moving from a dead stop.

Practice on a hill that is not too steep. Try to find a hill the does not have too much traffic. Then you don't have to worry about rolling back into another car.

Here are some suggestions for starting on a hill:

1. Because you have stopped, your left foot should be on the clutch. Your right foot should be on the brake pedal.

2. Shift into first gear.

3. When you are ready to go and when the way is clear, push down lightly on the gas pedal.

4. Let up the clutch.

Do steps 3 and 4 smoothly, but as quickly as possible after you take your foot off the brake. Otherwise, you will roll back a lot.

Using the emergency brake when starting on a hill

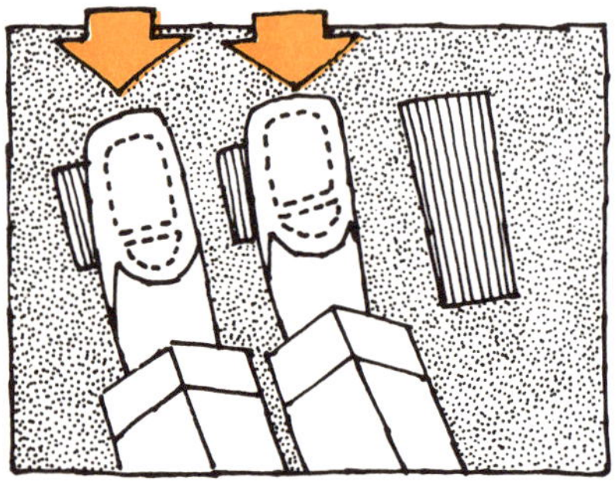

1 Stopped: Feet on clutch and brake.

2 Put on parking brake.

3 Shift to first.

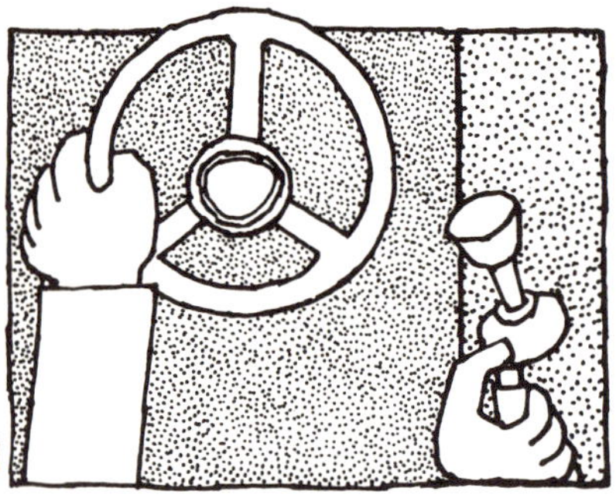

4 Left foot on clutch, right foot on gas pedal. Release emergency brake.

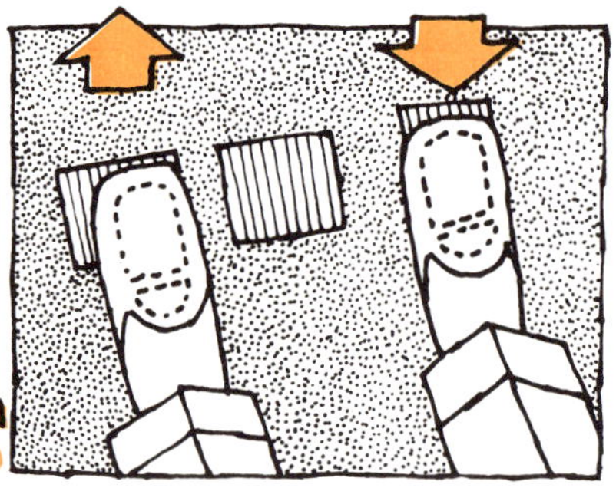

5 As parking brake is released: Push down on gas. Release the clutch.

6 You're on your way!

Steering, turning, backing 9

Learning to handle the car takes practice. But soon, you will be turning, steering, parking, and all the rest.

Steering

Pretend that the steering wheel is a clock. Place your left hand where the 10 would be on a clock. Place your right hand where the 2 would be.

This 10-and-2 position is probably the safest position for handling and controlling the car. This position makes starting to turn easy.

Steering on a straight road

The car will not go in a straight course on its own. It will not stay in one lane if you only hold the steering wheel. The car wanders. It wanders because of a bump in the road or the slope of the road. Some cars wander a lot because of problems with the tires or steering parts.

As the car moves off course, you must turn the wheel slightly to bring it back. You must always be steering to correct the wandering and to keep the car on course.

A good position for your hands while driving

Turning

Use **hand-over-hand** steering for turning corners. Use it for any turns that call for more than a half a turn of the wheel.

In hand-over-hand turning, one hand crosses over the other to turn the wheel. To make a right turn:

1. Start with your right hand at the top of the wheel.

2. Turn until that right hand is near the bottom of the wheel. The wheel should just slide through your left hand.

3. Keep turning with your left hand now. Also, bring your right hand back to the top of the wheel.

4. Keep turning until the turn is completed.

For a left turn, start with the left hand at the top of the wheel.

After you finish the turn, the steering wheel should return to a straight position. It should return by itself. Let the wheel turn through your hands. Just rest your hands lightly on the wheel. Then you can control the speed at which the wheel returns to a straight position.

If you were moving slowly, the wheel may not straighten out by itself. Then you must straighten it out.

This move takes a little practice. You have to learn not to oversteer when you straighten out. You could end up in the wrong lane or on the curb.

If your car has power steering, it will turn easily. You do not have to work very hard to make the car turn. And just a slight turn of the wheel could make the car swerve.

Before any turn:

– Signal at least half a block before you turn.
– Slow down.
– Check traffic and watch for people on foot.

For better hold on the road, slow down before you make the turn. Then, give the engine a little gas as you come out of the turn.

Making a right turn. To make a right turn, you should be in the far right-hand lane. If you are not in that lane already, get ready to move into it. Check traffic. Give a right-turn signal. Move into the right-hand lane when it is safe to do so. Don't get in the way of traffic already in that lane. Don't cut in too close to another car.

Signal for the right turn at least a half a block before you turn. Start turning when the front wheels are about even with the corner of the curb. Starting the turn too late might land you in the left lane after the turn. You would be in front of oncoming traffic. Starting the turn too soon could land you on the curb.

Many persons make the mistake of swinging out to the left just before making a right turn. That kind of move confuses drivers behind you.

Follow the turn in the road with the turn of your wheel. Use the hand-over-hand method. Once you are around the corner, let the steering wheel straighten out. Or straighten it out yourself, if needed.

Making a left turn. Make a left-hand turn from the lane closest to the center of the road. Signal

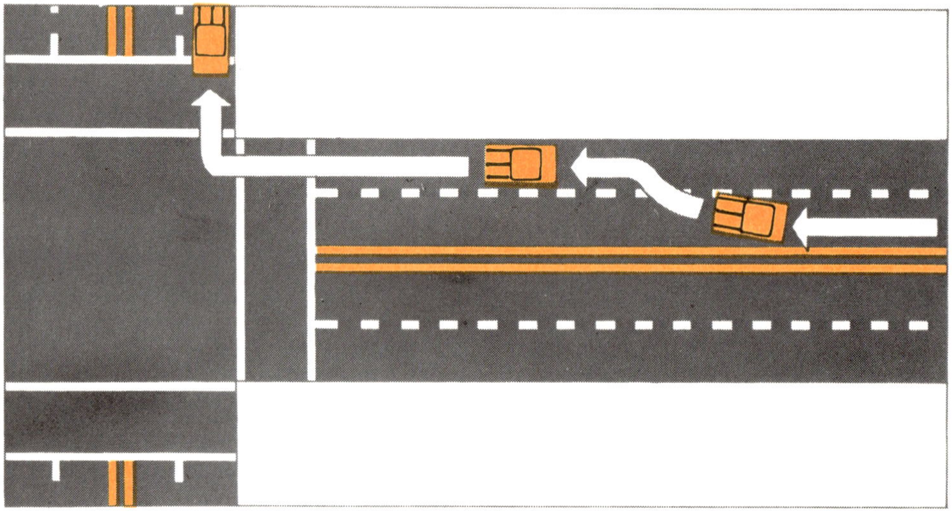

RIGHT TURN

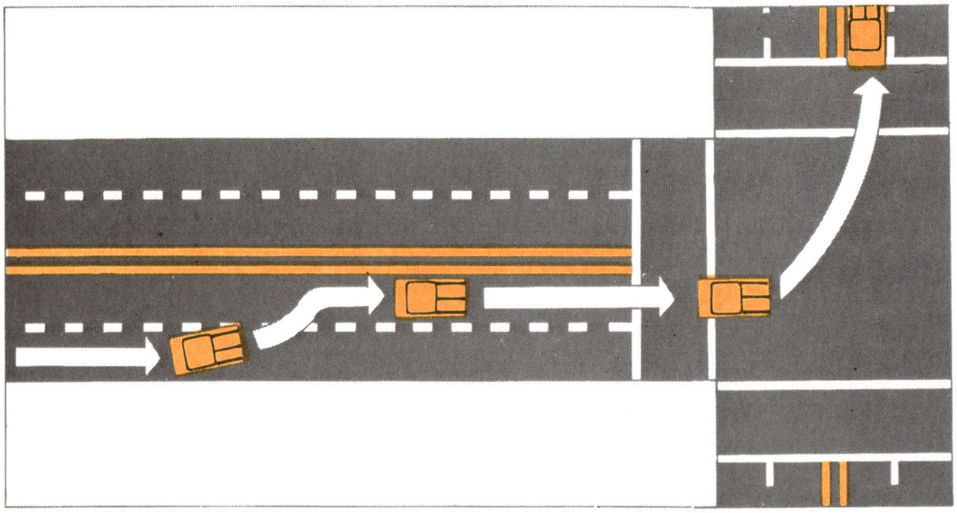

LEFT TURN

Is the way clear?

Get out and check before backing.

left for a lane change if you are not in the correct lane. Check traffic. Get into the turning lane when it is safe to do so. Change lanes at least a block ahead of your turn.

Let's say you're at the point where you want to turn left. Keep signalling with your hand or lights that you are going to turn left. Wait until there is an opening in the oncoming traffic. Then begin your turn into the street you now want to be on. Steer the car into the lane closest to the center of the road on that street. Use the hand-over-hand method to make the turn. Let the wheel straighten out when you have finished the turn.

With a left turn, you must cross a lane or more of oncoming traffic. Make sure you have plenty of time to get across those lanes. You must yield the right of way to oncoming traffic.

If oncoming traffic keeps you from turning, wait in your lane in the middle of the intersection. But keep the wheels of the car straight. Then if you are hit from behind, you won't be pushed into the path of oncoming traffic.

If you're making a left turn from a one-way street, turn from the left lane.

Backing up

Before you back out of your driveway, check behind the car. Walk around the car to see if there are any children, pets, or toys nearby.

When you back up, look over your shoulder. Don't just look in the rearview mirror. There are too many blind spots.

Back up slowly. Keep looking behind you. Check all directions. When you turn the wheel, look over your shoulder in the direction in which you are moving.

Even experienced drivers sometimes get confused about which way to steer when backing up. Turn the steering wheel in the direction you want the back of the car to go. If you turn the wheel to the right, the back of the car goes to the right. The front swings out to the left.

If you turn the wheel to the left, the rear of the car goes to the left. The front of the car swings out to the right.

Practice backing up. Practice in a place where

Turn steering wheel to the right: the car rear goes to the right and the front swerves to the left.

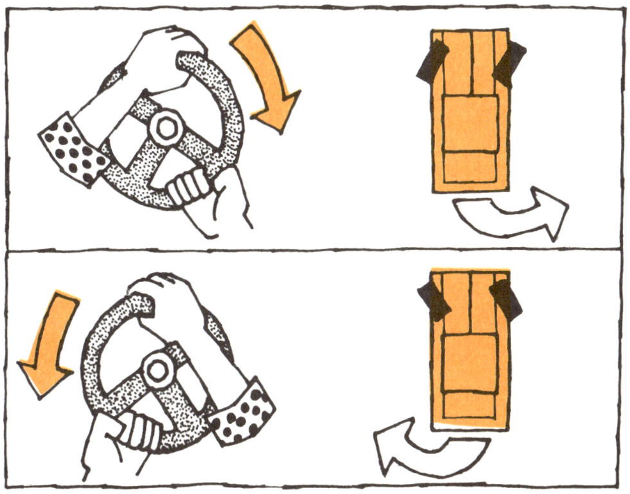

Turn steering wheel to the left: the car rear goes to the left and the front swerves to the right.

Stop here to look for people on the sidewalk.

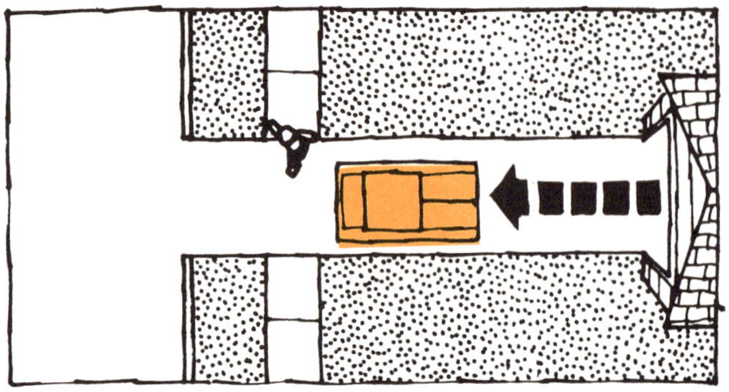

Stop here to look for traffic on the street.

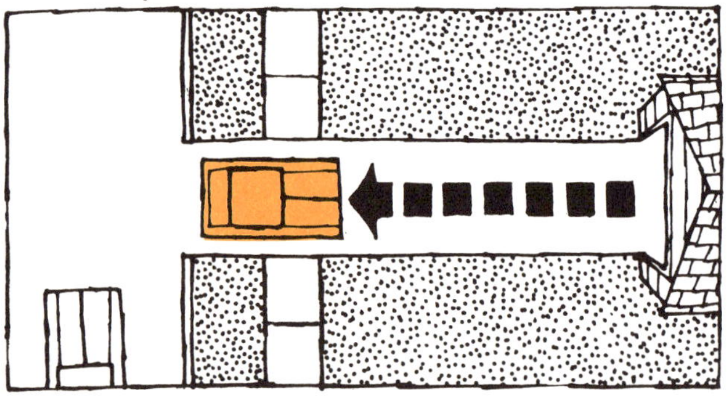

there are no other cars or pedestrians. With a little practice, the steering begins to make sense.

To back into a street from your driveway, follow these safety rules:

1. Make sure the wheels are straight before you back up. Pull forward to straighten them if they are not straight.
2. Check behind you as you back up.
3. Stop before you reach the sidewalk. Make sure no pedestrians are crossing behind you.
4. Stop before you enter the street. Wait until you have a good opening before entering traffic. Traffic on the street has the right of way. You must wait.
5. Back into the lane you want to be in.
6. Don't face front until you have brought the car to a full stop.
7. Don't back into a highway or into heavy traffic.

Suppose you want to back into the far lane. You will have to wait until the road is clear in both directions. This could mean a long wait. Backing into the far lane might be too dangerous.

It is a good idea to back into the near lane instead. You will not be facing the direction you want to be going in. But you only have to drive around the block.

U-turn

When you make a U-turn, you turn around in one move. You end up facing in the opposite way. The turn looks like the letter U.

Make U-turns only where they are legal. They are not legal on highways. They are not legal if you have to cross a double solid line. They are not legal if a sign says *No U-turn.*

When you make a U-turn, choose a safe place. Don't make a U-turn near an intersection, a curve, or the crest of a hill. Make sure there are no children, pedestrians, or other vehicles nearby. Wait until there is a long enough break in traffic. You need extra time in case you misjudge the turn and don't make it in one movement. Then you will have to back up before you can go forward again. Allow time for extra movements that may be needed.

Follow these steps to make a U-turn:

1. Find a safe spot where you can easily see traffic coming from both directions.
2. Pull close to the right-hand curb.
3. Signal for a left turn.

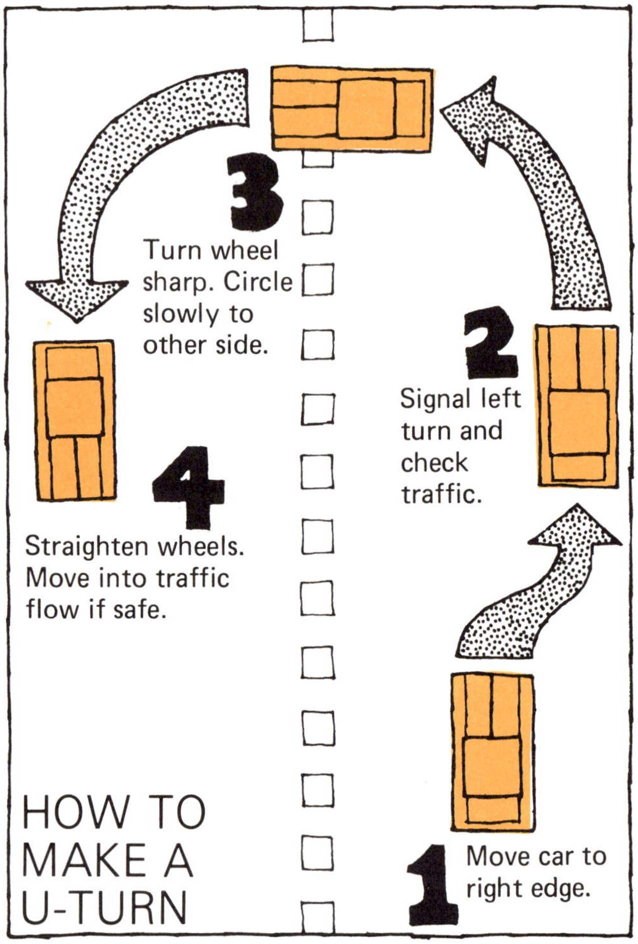

4. Check for traffic in both directions. Look over your left shoulder just before you start your turn.

5. Turn the steering wheel quickly to the left. Use the hand-over-hand movement. Move the car slowly, though.

6. Turn until the car is facing in the opposite way from where you started.

7. If you misjudged, and didn't make the complete turn, then you must make a 3-point turn (see next section).

8. Straighten your car's wheels. Move into the traffic flow if safe to do so.

3-point turn

If the street is too narrow for a U-turn, you will have to make a 3-point turn. In a U-turn, you make the turn in one movement. In a 3-point turn, there are three movements. You go forward, then in reverse, then forward again.

Do not make a 3-point turn near an intersection, a curve, or the crest of a hill. Do not make the turn where you can't see well all around. Pick a spot where there are no trees, or poles, or other objects that you could back into.

To make a 3-point turn, follow these steps:

1. Pull over to the right-hand curb.
2. Check traffic in both directions. Wait for a large enough opening in traffic to give you time to make the turn.
3. Signal for a left-hand turn.
4. Turn the steering wheel quickly to the left as you move slowly forward.
5. You should be facing the curb on the other side of the street. Stop a few feet from the curb.
6. Shift into reverse.
7. Turn the steering wheel to the right as you slowly back up.

8. When the rear of the car is close to the other curb, stop.
9. Shift to a forward gear. Check traffic.
10. Turn the steering wheel to the left as you move forward to complete the turn. You should now be facing in the opposite way from your direction at the start.
11. If you did not complete the turn, repeat steps 6 through 10. (Back up and then forward again.)

10 Parking the car

There are two types of parking—angle and parallel. Angle parking is used in parking lots and parking garages. It is also used on some streets. Parallel parking is parking next to the curb on the street. It is harder to do than angle parking.

Practice both types of parking. But practice away from other cars. Try parking between lines or markers.

Angle parking

Drive slowly in parking lots and garages. Watch out for other cars backing out. They may not see you coming. Honk your horn lightly if needed.

To park at an angle, follow these steps:

1. When you see a parking spot on the right, signal for a right-hand turn. Stay about six feet from the row of cars on the right.
2. Turn sharply to the right. But move the car slowly. Slowly guide the car between the lines of the parking space.
3. Once you are in the space, straighten out the wheels of your car. Do this as you inch forward with your foot on the brake pedal.
4. There will be a curb or marker to show how far you should pull in. Stop a few inches before the tire reaches the curb or the front end reaches the marker.
5. Put the gearshift lever on park if you have a car with automatic transmission. On a car with standard transmission, keep it in reverse. Set the parking brake. Get out of the car. Lock all doors.

Backing out

You may have trouble seeing as you back out. So move very slowly. To back out of a parking spot, follow these steps:

Stay about six feet from the row of cars before turning.

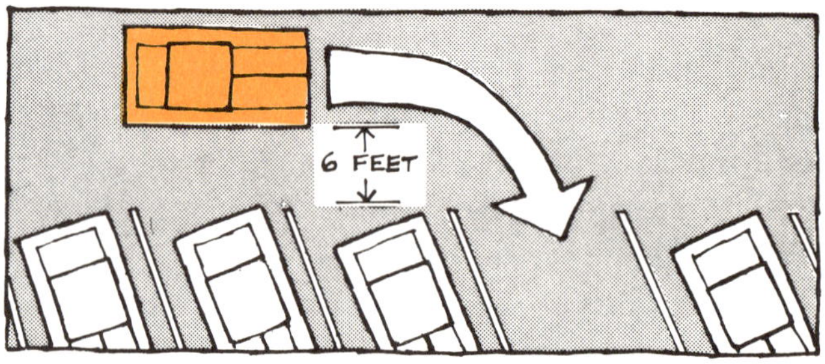

Avoid hitting the car beside you as you back out. Don't turn while your car is still alongside another car.

1. Look over your shoulder as you slowly back straight out. At all times watch for people on foot and for other cars. Back straight out until you can see to the left and right behind you.
2. Stop at this point and look around. If the way is clear, start backing out again. Keep looking over your shoulder, especially in the direction you will be turning.
3. Back straight out. Turn only when you are sure the front of your car will not hit the car next to yours.
4. When clear, turn the wheel and swing out into the proper lane.
5. Shift to a forward gear. Drive slowly if you're in a parking lot or garage. Watch out for other cars backing out. Parked cars block your view. So watch out for cars turning down your lane. Other drivers' views are blocked, too.

It's hard to see traffic as you back out. Go easy.

Parallel parking

Parallel parking usually calls for fitting your car into a very small space. It takes a lot of time and sweat to do this until you learn how.

In many states, you must be able to parallel park to pass your driver's test. So it is a skill you must know.

A parking spot should be at least five feet longer than your car. You need that extra space for getting in and out. When you find a parking space, take these steps:

1. Signal that you are stopping. Put on your right-turn signal.
2. Pull up past the parking spot. Stop next to the car in front of the empty space. The rear bumpers of the two cars should be even. Your car should be about two feet away from the other car.
3. Shift into reverse. Check traffic behind you. Make sure you will not be backing into traffic moving in your lane.
4. Turn the steering wheel to the right. Slowly move backward into the parking space. Stop when your steering wheel is even with the rear bumper of the car in front. Your car should now be at a 45-degree angle with the curb.
5. Stay in reverse. As you slowly back up, straighten out your front wheels. Keep backing up until the front of your car is even with the rear of the car in front.
6. As you back up slowly, turn the steering wheel sharply to the left. Watch out for the car parked behind you. You should now be in the parking space and parallel to the curb.
7. Stop before you hit the car behind yours. Shift to a forward gear. Straighten out your car's front wheels as you slowly move forward. Leave space between your car and the cars in front and back of it.

Your car should be within about one foot from the curb. The legal distance could be more than a foot or less than a foot. The distance varies from state to state.

So now the car is in the space. But you may have to move it closer or farther from the curb. You may have to back up or go forward a little as you turn the wheel. But once you get the hang of it, you shouldn't have to do too much adjusting.

8. Put the gearshift lever into park (or reverse on a standard transmission car). Set the parking brake. Get out of the car. Lock the car doors. Put money in the parking meter if needed.

Uphill and against the curb

Downhill and against the curb

Uphill without a curb

Getting out of parallel parking space

Follow these steps to get out of the parking space:

1. Release the parking brake.
2. Shift into reverse.
3. Back up until your car almost touches the car behind yours.
4. Shift to a forward gear. Signal left. Check traffic. Use mirrors and look over your shoulder at the last minute.
5. As you move slowly forward, turn the steering wheel sharply to the left.
6. Pause when the steering wheel is even with the rear of the car in front. Start moving the steering wheel to the right. Don't turn so sharply to the right that you sideswipe the car in front. Go forward until the wheels are in line for you to move into the flow of traffic.
7. Pull into the flow of traffic when there is a safe opening.

Parking on a hill

When you park on a hill, make sure the car will not roll into the street. If you are parallel parking, park close to the curb.

When the car faces uphill, turn the wheels away from the curb. The tires should lightly touch the curb.

When the car faces downhill, turn the wheels into the curb. The tires should lightly touch the curb.

If there is no curb, always turn the wheels to the right. This is the rule whether you are facing up hill or down hill.

When parking on a hill, be sure to set the parking brake.

11

Driving on the open road

The open road does not have the many stop signs and traffic lights you find in the city. The open road is the network of roads and highways outside of the city.

The open road includes many kinds of roads. Many open roads are for steady-speed traffic. There are few stops. Freeways are open roads. (We will look at them in the next chapter.) County roads are open roads. The interstate routes are open roads. There are also many back roads. The back roads may not be paved. They are not designed for high-speed travel.

The speed limits on the open road vary. On a back road, the speed limit may be 35 or 45 miles per hour (mph). On the freeway, the limit is 55 mph. That is the national speed limit.

Following distance

You should keep a safe distance from the car in front of you. If the car in front of you stops quickly, you should be able to stop without hitting it.

How can you be sure you are far enough from the car in front of you? Some people stay at least one car length behind for every 10 mph they are

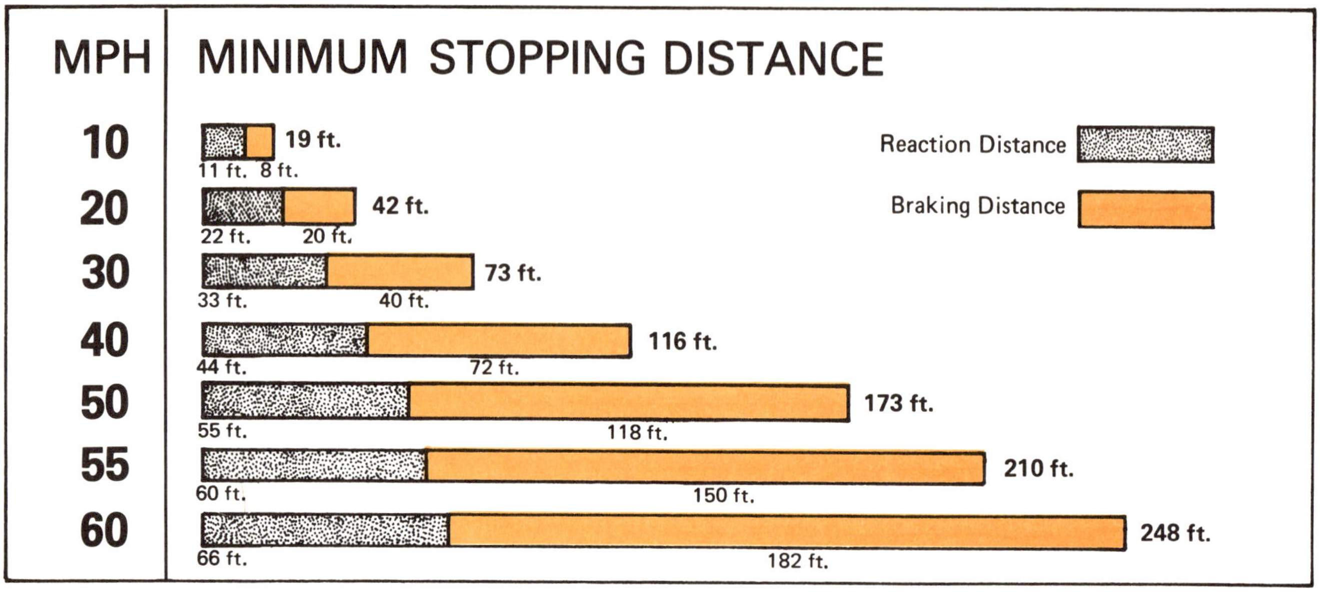

driving. When traveling 45 mph or faster, they stay two car lengths behind for every 10 mph.

Here's another way to keep a safe distance from the car ahead: *Stay at least two seconds behind the car in front of you!*

Here's how to know if you're two seconds behind the car ahead: Watch when the car ahead passes an object beside the road. Let's say the object is a road sign. Begin counting at the moment the car passes the sign. (Count *one thousand one, one thousand two.* That takes about two seconds.) It should take at least two seconds before your car passes the same sign that the car ahead just passed.

In bad weather, you should stay even farther behind. You need two or three times more room to stop. Also, stay farther behind when driving downhill.

How do you *feel* about road hogs? What do you *do* about road hogs? It's best not to *do* what you *feel*.

Why is it so important to keep good distance between cars?

Suppose you are driving at 55 mph. The driver ahead of you starts to brake. That car needs about 150 feet to stop. But you'll need at least 210 feet.

You must allow room for reaction time. It takes you a moment to realize the car ahead has slowed down. It takes you a moment to realize you must put your foot on the brake pedal. So even if you react quickly, you will travel at least 60 feet before you hit the brake.

If you are driving too closely behind the other car, you won't have enough room to stop safely.

If the road is wet or slippery, you will need about three times more room to stop. And if your brakes or tires are in poor condition, you will need even more room to stop.

Oncoming traffic

Oncoming traffic is traffic coming from the opposite direction. You are in your lane. The oncoming cars are in their lane. You simply drive by one another. This is an easy enough move—usually.

Every so often you meet the road hog. He drives too close to the center line. He may even be a little over the line and partly in your lane.

A road hog may just be a very tired driver. Or maybe he is a driver who misjudged a curve and ended up over the center line.

Whatever the reason, there the road hog is. He is "taking his half of the road from the middle." The last thing you should do is get angry and try to get back at him. Keep your temper. Just pull over to the right a foot or two. Give the oncoming car extra space to get by.

Hills

Hills on two-lane highways can be very dangerous. You cannot see over the crest of the hill as you drive uphill. You don't know what is on the other side of the hill. There may be a vehicle approaching in your lane. There may be a slow-moving vehicle in your lane.

Don't pass on a hill unless there is a passing lane for uphill traffic. Slow-moving traffic should move to the right-hand lane. Faster traffic can pass in the lane to the left. But, if you are on a two-lane road with no passing lane uphill, don't pass. Even if the car ahead of you is moving very slowly, don't pass. You don't know what is on the other side of the hill.

As you come to the crest of a hill, let up on the gas pedal. Slow down a little. There might be trouble in your lane on the other side of the hill.

There is another reason to slow down now, too. You will pick up speed as you go down the hill.

As you drive down the hill, take your foot off the gas. You may want to shift into a low gear, too. Then the engine will slow you down even

You never know what you may meet beyond the crest of the hill. So slow down and get ready for anything.

Stopping distance when going downhill is much longer.

more. You may also have to pump or press the brake lightly. But don't put a lot of pressure on the brake pedal. Too much heat could build up in the brakes or the tires.

Don't coast downhill in neutral. Coasting is driving in neutral. Your foot is off the gas pedal. You just steer.

Coasting is not legal in most states. It is dangerous. It is bad for the car.

When you coast, the engine is not helping to slow down the car. The car can build up a lot of speed. You could burn out the brake linings if you had to stop while speeding down the hill.

Downhill passing. Suppose you are following a slow-moving vehicle. You can't pass because you are climbing uphill. You can't see over the crest. But when you get to the crest, you see the road is clear for passing.

Before you start to pass, wait a moment. You may find you don't need to pass after all.

Moving uphill, the vehicle in front of you was working hard. It had to move slowly. But going downhill, that slow vehicle picks up speed. It may pick up so much speed that you will have trouble getting past it. It may go the same speed you are going or faster.

So let a slow vehicle reach its downhill speed before you decide whether to pass it.

Following distance on a hill. When you are driving downhill, keep a good distance from cars in front of you.

You need more room to stop on a hill than on a

flat road. This is because gravity pulls the car downhill.

To be safe, stay at least two car lengths for every ten miles per hour.

Curves

You can't see around a curve. Trees or boulders or hills or buildings cut down your view even more. You can't see if there is any problem in the road around the curve.

Slow down before you enter a curve. If there is a posted speed limit for the curve, obey the speed limit. Slow down to that speed before you enter the curve.

You shouldn't have to brake when you are rounding the curve. For best traction, you should be feeding gas to the engine. You should be pushing the gas pedal gently.

If you misjudged the curve, you may have to brake. Use your brakes if you feel you are going around the curve too fast. If you round the curve too fast, you could lose control of the car. But brake gently. Don't slam on the brakes.

Stay in your own lane as you go around a curve. When you're curving to the right, your car pulls toward the center of the road. Don't let the car stray to the center of the road or into a lane for oncoming traffic.

When you curve to the left, the car pulls toward the edge of the road. But at safe speeds, you can

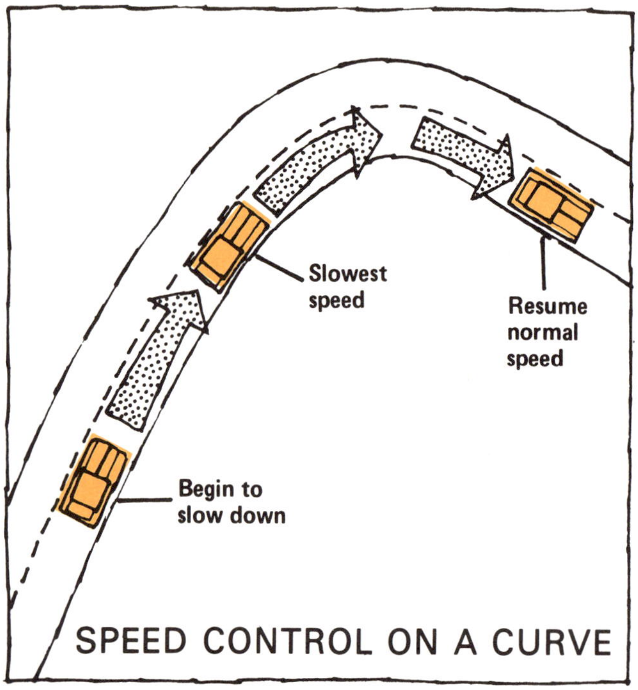

SPEED CONTROL ON A CURVE

ROAD DESIGN MAKES A DIFFERENCE

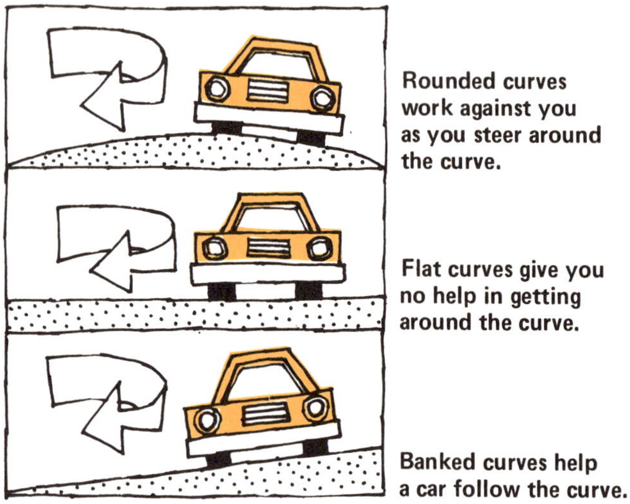

prevent the car from flying off the road. Stay as close to the middle of your own lane as you can.

The laws of nature keep the car going straight unless you steer it round the curve. The faster you go, the harder it is to control the car. The sharper the curve, the more trouble you will have controlling the car. So you must drive very slowly on sharp curves.

You must also go slow on curves for another reason. You can see only a short distance ahead of you. You should be able to stop within the distance you can see. Suppose you can see only 100 feet ahead of you. Then you should be able to stop within 100 feet.

Your view of the road won't improve until you get around the sharpest point of the curve. Don't speed up until you can see clearly ahead of you. Don't start to speed up until you are coming out of the curve. Then speed up gradually.

Railroad crossing

Railroad crossings are clearly marked. There is a round warning sign before the crossing. The sign is yellow and black. At the track, there is a criss-crossed sign. Some crossings also have a gate that comes down when a train is coming. There also may be flashing lights when a train is coming.

When the lights are flashing, stop. Don't try to beat the train. Wait for the train to pass. There is always the chance that your car could stall on the tracks. A train wouldn't be able to stop for you.

When you come to a railroad crossing, slow down. Turn down your radio. Listen for a train

Let other cars cross the tracks before you start across. Don't get caught on the tracks behind a stalled car.

whistle. Look up and down the tracks. If the way is clear, go across.

Don't drive across tracks until the car in front of you is all the way across. You could be trapped on the tracks if a car ahead of you stalled.

Don't drive across the tracks right after the train has passed. Wait a few seconds to be sure a second train isn't coming.

Buses and many trucks must come to a full stop before driving over railroad tracks. A bus driver opens the door of the bus to listen for train whistles. If you are driving behind a truck or bus, be ready to stop at the railroad crossing.

Remember that when the tracks are wet, they are slippery. In rain and snow, drive very carefully over the tracks so you do not skid.

Soft shoulder

The shoulder of the road is the side of the road. Sometimes it is not paved. It may be soft. There are often signs warning about soft shoulders. If your right wheels are on the soft shoulder, the car will pull hard to the right. The car will be hard to steer and control.

Don't drive on the shoulder unless you must. For example, you may be forced off the road. Or you may have to stop for an emergency.

If you are on the shoulder, you will need both hands for steering. Don't brake. Let up on the gas. Slow down enough so that the car won't be thrown when you steer back onto the road. Steer back onto the road when the way is clear. Don't step on the gas until you are completely on the road again.

Passed by another car

As a beginning driver, you may be passed by many cars. Don't speed up when someone is passing you. Don't move toward the left lane. Stay to the right. Keep your same speed or slow down a little. By slowing down, you help the other driver get past you sooner. Give the other driver the chance to get back into the right lane as soon as possible. This is very important on a two-lane road.

If you are passed by a large truck or bus, hold on tight. The fast-moving truck or bus can cause a vacuum (or wind) that pulls at your car.

Passing another vehicle

Passing another car on a two-lane highway is a tricky move. A two-lane highway has one lane of traffic for each direction. To pass, you must use the lane that oncoming traffic moves in.

Passing on a two-lane highway calls for good timing and judgment. But here are some pointers to help you.

DRIVING ON THE SHOULDER

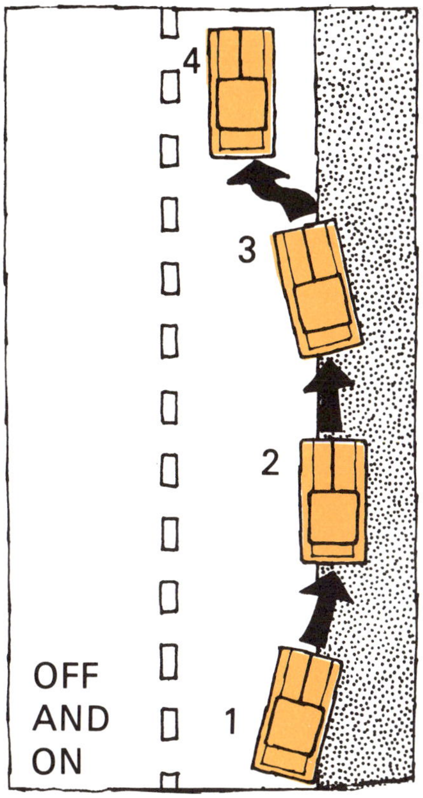

Straighten car quickly when back on road.

Turn back on to the road. Hold wheel firmly.

Check behind and ahead to see if it's safe to get back on the road.

Slow down.

Is it legal to pass here? You can pass only in areas where the law allows passing. Passing is not legal in areas where it is not safe. If you cannot pass, there is usually a sign that says *No Passing.*

It is not safe or legal to pass:

- on a hill (unless there is a passing lane)
- on a curve
- at an intersection or railroad crossing
- if there is a double solid line in the center of the road
- if there is a solid yellow line on your side of the center lines
- if a sign says *No Passing.*

You can pass if the center line on your side is a broken white or a broken yellow line. But you must be sure it is safe to pass.

When is it safe to pass? You must use your good judgment to answer this question. To decide whether to pass, you must weigh these factors:

- your speed
- the speed of the car you want to pass
- the time it will take you to pass and get back into the right lane
- the pick-up your car has. (That is the reserve power that allows your car to pick up speed.)
- the distance between your car and any oncoming car
- the speed of an oncoming car
- the distance between the car you are passing and the car ahead of that car. (Is there enough room for you to get in between them without getting in their way?)
- your closeness to a **No Passing** area.

Don't try to pass unless you are sure you have enough time and clear highway ahead of you. If you have any doubt about that, don't try to pass. Play it safe.

How to pass. Make sure you are in a legal passing zone. Make sure the zone doesn't end before you can finish passing. Make sure there is no curve or hill just ahead. The curve or hill would keep you from seeing oncoming traffic in the left lane. Think about the points in the above section.

If you decide you can pass safely, follow these steps:

1. Look in your rearview mirrors. Make sure no other car behind you has pulled into the left lane.
2. Check ahead of you once more. Make sure no oncoming car is too close.
3. Before you start to pull into the left lane, put on your left signal light. Or give a hand signal.
4. Pass from a safe distance. Begin pulling into the left lane several car-lengths before you reach the car you are passing. In this way, the car ahead won't be blocking your view of the left and right lanes. And you will have enough room for picking up speed. You will also have enough room to move back into the right-hand lane if you see an oncoming car.
5. Tell the driver of the car ahead of you that you are passing. Sound your horn as you start to pass. At night, flash your lights from dim to bright, then back to dim.
6. Pick up speed quickly to pass. But don't go faster than the speed limit. You should be going five to 10 mph faster than the car you are passing.
7. To see if it is safe to get into the right lane again, look in the rearview mirror. It is safe when you can see in your mirror the front of the car you passed.

   (At high speeds, wait until you can see a few car lengths of road in front of the car you passed.)
8. Signal that you are going to move into the right-hand lane.
9. After you are in the right-hand lane again, turn off your signal light.

Freeway driving 12

When someone says the word *freeway*, what's the first word that comes to mind? For many persons, that first word is *fast.*

Freeway driving is fast compared to driving on other roads. You must drive at a fast, steady pace of 55 mph, unless road or weather conditions are bad.

Freeways are also called expressways, turnpikes, thruways, or parkways. Such roads have these features:

– There are no intersections.
– There should be no pedestrians.
– There are no traffic lights or stop signs.
– The road is divided. There is a strip of land that divides the traffic moving in opposite directions.
– There are exit and entrance ramps.
– Traffic moves at a steady, fast pace.
– There is more than one lane in each direction.

Freeways can be a lot safer for driving than city streets because of their features. But those features can also make freeway driving dangerous. Drivers figure the road is wide open and straight. So they are tempted to drive too fast. But any hazard becomes even more dangerous if a car is moving too fast.

Remember a few basic pointers when driving on the freeway:

1. Don't make any abrupt or sudden moves.
2. Signal when you change lanes.
3. Keep close track of traffic around you.
4. Keep a safe following distance. Keep a bit more than one car length behind for every 10 mph you are moving. Try to keep this spacing even in heavy freeway traffic.

Now let's picture ourselves riding on a freeway–from entrance to exit. You will probably see that freeway driving is easier than city driving. You will also see where the dangers may lie.

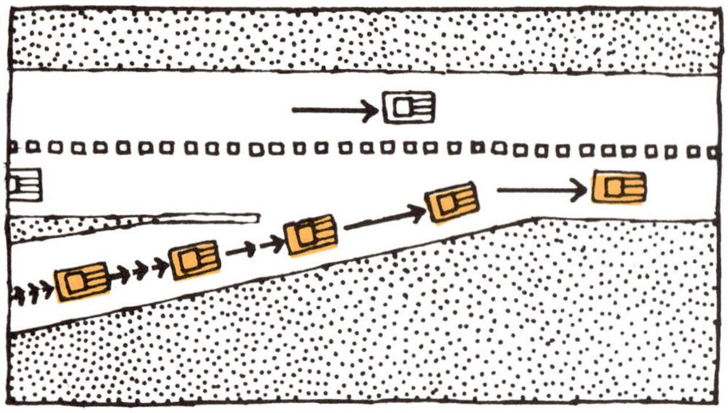

Entering the freeway

To enter the freeway, you must drive on a ramp or entrance. The entrance may be a long curved roadway. The speed limit is lower than the speed limit on the freeway.

There is usually an acceleration lane. The acceleration lane runs alongside the freeway for a short distance.

The acceleration lane is supposed to give you a chance to build up your speed. You don't have to enter 55 mph traffic from a full stop. Once you have built up your speed, you can blend into the traffic flow more easily and safely.

If traffic on the freeway is heavy, you may have to wait for an opening. In this case, slow down or stop at the beginning of the acceleration lane. Then you will still have enough space left to build up your speed when you see an opening coming.

Suppose there is no acceleration lane. Or suppose the acceleration lane is very short. Wait at the entrance of the freeway until there is an opening in traffic. Wait for a large opening. Then you will have plenty of time to speed up. Traffic should not have to slow down for you. So you must build up speed quickly to match the speed of other cars on the freeway.

To enter the traffic flow, check the traffic on the freeway. Signal. Build up your speed. Merge into the lane closest to you when it is safe to do so.

Traffic already on the freeway has the right of way. But other drivers may try to help you enter.

Many drivers move into a left-hand lane so an entering vehicle can get into the traffic flow. If the other drivers do not move over, the entering driver will have to wait. He may have to wait at the end of the acceleration lane. Then he must enter from a stop or a very slow speed. That can be dangerous.

When you are on the freeway, help out an entering vehicle. If you see a sign warning about merging traffic, try to pull over to a middle or left-hand lane. Make room in the right lane for entering vehicles.

The acceleration lane isn't always long enough. Let's say a driver comes to the end of the acceleration lane. And let's say he finds no opening in the traffic to fit into. Then there is danger for many persons. There is danger for traffic on the freeway and for the entering driver. There is also danger for other drivers behind him trying to enter the traffic flow.

Moving along

Drive in the right-hand lane except to pass or make room for entering traffic.

Keep up your speed on the highway. A slow-moving car can be bumped in the rear. Or it could be side-swiped by a car moving up fast from behind. The driver behind it may have to change lanes too quickly to avoid the slow-moving car.

Try not to drive more than five mph slower than the speed limit. Or don't drive slower than any posted minimum (slowest) speed. You can get a ticket if you drive slower than the posted minimum speed. But if road or weather conditions are poor, drive at a speed that is safe for those conditions. That may mean driving much more slowly than the posted minimum speed.

If you must drive very slowly, put on the four-way flashers to warn drivers behind you.

Don't drive faster than the posted speed limit. Check your speed regularly. Learn to glance quickly at your speedometer.

You must be very alert when you drive on the freeway. Cars move at a fast pace. The traffic situation can change quickly. One moment, you see no one in your rearview mirror. The next moment, someone is passing you.

Cars change lanes often. They pass often. You must watch for these sudden changes. Check your mirrors about every 10 seconds.

Passing on the freeway

If you want to pass, make sure no car or truck or bus is moving up fast behind you. If a vehicle is moving up fast behind you, don't pull into the left lane. The fast vehicle will probably move into it to pass you. Let it pass you before you move into the left lane.

If a car is moving up in the passing lane, let it go by before you move into the left lane.

When you are going to pass, look in the rearview mirrors. If the passing lane is clear, signal left. Pull into the passing lane. After you have passed the car in front, look in your rearview mirror. When you can see a few car lengths of road in front of the car you passed, you can move into the right lane. Put on the right-turn signal. Move smoothly into the right lane. Turn off your signal light.

Don't cut back into the right lane too soon. Don't force the car you passed to brake because you have cut in. You are not on a two-lane road. You do not have to worry about oncoming traffic.

If a car passes you and cuts back in too soon, slow down a little. Help to set a safe following distance again. The driver behind you should see that he is gaining on you. He will slow down a little, too. Or he will pass you.

Leaving the freeway

Signs tell you when you are coming to an exit. They give you plenty of warning that your exit is coming up. They give you time to get ready.

Get in the proper lane about a mile before the exit. Most exits are on the right. So you should probably be in the right-hand lane. (If the exit is on the left, get in the left lane.) If traffic is heavy, try to get into the proper lane when you see the two-mile warning for the exit. Don't wait until the last minute to try to squeeze into the proper lane.

As you approach the exit, there may be a slow-down lane. This lane gives you a chance to slow down gradually before you reach the exit ramp. The speed limit on the exit ramp is much slower than on the freeway.

Superhighway driving isn't always free.

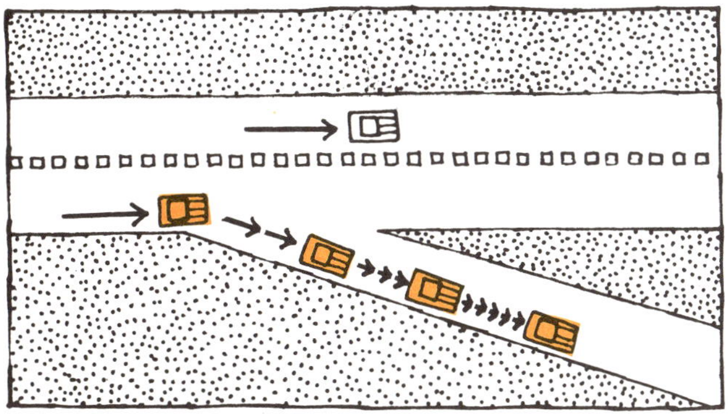

Signal that you are going to exit before you reach the exit.

If there is a slow-down lane, keep up your speed until you reach the slow-down lane. Then gradually lower your speed.

If there is no slow-down lane, start to slow down before the exit. Slow down more when you are off the freeway.

If you miss your freeway exit, never try to back up. This is not only against the law, it is dangerous! Instead, drive to the next exit.

Check your speed as you come off the freeway. You have been moving fast on the freeway. You will not be able to judge your speed very well when you get off. Anything below freeway speeds will feel like a crawl. You may *think* you are moving slowly. But you may be moving too fast for the exit ramp or for city driving.

Toll roads

If you were on a toll road, you will have to pay for using the road. The toll booth is at the end of the exit ramp. Slow down as you approach the toll booth.

Don't fumble in your pockets or purse for money while you are driving. If you know you will be on a toll road, put some money in a handy place before you start out. Then you can simply reach for the money when you have stopped at the toll booth. You will save time. You won't delay others behind you, either.

13 Driving in the city

City driving is stop-and-go driving. You must stop often for red lights, stop signs, and pedestrians.

You need to be very alert when you drive in the city. And you may need to be very patient, too. You probably won't be able to move as quickly and as steadily as you would like. Some days, it may seem as if you are hitting every red light in the city.

But you need to be patient. When drivers become angry and lose patience with traffic, they make mistakes. Then accidents happen.

Speed

The speed limit is 30 mph on most city streets. In school zones, the speed limit may be only 20 mph. Learn the speed limit in your city. Obey that speed limit unless other speeds are posted.

Keep pace with the rest of the traffic. If you are passing many other cars, you are probably driving too fast. If you are being passed a lot, you may be driving too slowly.

Driving much more slowly or much faster than the rest of the traffic causes accidents. Heavy traffic may be moving at a slow pace. It may be moving more slowly than the speed limit allows. But if you are driving at the same pace as the traffic you are in, you are driving at a safe speed. Of course, you'll never want to go faster than the speed limit even if most cars around you are doing so.

Following distance

Keep enough distance between your car and the car in front of you. This rule still stands–one car length of distance for every 10 miles per hour.

This may seem hard to do. Other cars may slip in front of you if the following distance is too great. If someone does slip in, drop back if you can. Don't tailgate.

If someone tailgates you, let him pass if you can. Don't let a tailgater force you to drive too fast. Try to give him room to get by you.

Driving in your lane

As you get used to driving, you will get used to staying in your lane. You will get a feel for the width of your car. You will be able to keep the car between the lines of the lane. You will even be able to keep the car in the lane when the painted lines are worn away.

Getting to know which lane to drive in can save you time. And you can help keep traffic from jamming up behind you.

The right-hand lane is usually for slower traffic. The left lane is for faster traffic. But persons making left turns also drive in the left lane. If you are making a left turn, get into the left lane about 200 feet before the turn. In heavy traffic, get into the left lane sooner.

The right lane is usually for persons driving straight ahead or making a right turn. If you are going straight, or turning right, get into the right-hand lane.

When you are driving in the right-hand lane, watch out for parked cars. Be extra cautious. Someone may open the door of his parked car and step into traffic. Someone may step out between parked cars. A child may run out into the street.

Parked cars often mean surprise dangers. Drive carefully.

Sometimes, parked cars block the right-hand lane. Plan ahead when you drive in the city. Look ahead for parked cars blocking your lane. Get into the left lane in plenty of time. Then you won't have to change lanes or stop at the last minute.

Some streets may have three lanes going each way. The lane on the left may be for left turns only. The lane on the right may be for right-hand turns and going straight. The lane in the middle may be for going straight ahead, too.

Look for pavement markings that tell you what the lane is for. Big white arrows are painted in the lanes to tell you. They show left turn, right turn, straight, or straight and a turn. There also may be big white letters on the pavement. They say *Left Turn Only* or *Right Turn Only*.

Changing lanes

Signal when you plan to change lanes. Check behind you to make sure the lane is clear. Check to the side of you, too. If you are moving into the right-hand lane, glance over your shoulder first. If you have a mirror on the right side, then look in that mirror instead. If you are moving into the left lane, look into the left-hand mirror first. Move into the desired lane only when the lane is clear. Don't cut in front of another car.

Change lanes when you must make a turn or avoid a parked car. Don't change lanes often. Don't weave in and out of traffic just to get past cars you think are too slow.

Passing

Passing a car in city traffic can be tricky. There may be lots of traffic. There may not be much time or room for passing. Observe the rules for passing. Pay attention to lane markings and signs. Don't pass if you are not sure you have enough time or room.

In most cases, you should pass on the left. But you may pass on the right in two situations:

1. You may do it if there are two or more lanes of traffic moving in the same direction. Those lanes are moving separately from one another. You may drive past cars in lanes to your left.

2. You may also pass on the right if a car is stopped to make a left turn. But if you must go on to the shoulder to get past, wait until the car turns.

Always changing lanes doesn't get you there much faster. Besides, it's risky.

One-way street

A one-way street is usually clearly marked with a sign. The sign points in the allowed direction. If you find yourself driving the wrong way on a one-way street, find a driveway or a place off to the side. Get out of the way of oncoming cars. When traffic is clear, turn around.

You can drive on the left side of a one-way street. To make a left turn, turn from the left lane of the road. To make a right turn, turn from the right lane.

Common hazards of the city

Many city traffic situations can be dangerous unless you know how to handle them.

The law says stop when a school bus stops.

No law says you must stop for a city bus. But it pays to be careful when going around it.

School bus. When the red lights flash on the school bus, you must stop. You must stop whether you are behind the bus or driving towards the bus. The bus is loading or unloading students. You may not go again until the lights stop flashing or the driver or a traffic officer waves you on.

City bus. You do not have to stop when a city bus stops to load or unload riders. But you should slow down if you are passing a stopped city bus. Slow down if you are driving past in the opposite direction, too. Watch out for passengers getting off the bus. Children may run out in front of the bus when they get off. Watch for persons running to catch the bus, too.

Emergency vehicles. An emergency vehicle is an ambulance, a fire truck, or a police car. Sirens or flashing red or blue lights warn you that such a vehicle is coming your way. The vehicle is moving fast and must get somewhere in a hurry. When you see or hear an emergency vehicle, pull over to the side of the street. Wait for the vehicle to pass before you go again.

When you are not sure if the vehicle is on your street, pull over. Wait and see. Don't take chances. Stay out of the way.

Pedestrians. Intersections with traffic lights often have *Walk* and *Don't Walk* signals for pedestrians. But don't assume the pedestrian will obey the signals. A pedestrian may cross against the light. Or he or she might be in the middle of the road when the light changes. The pedestrian in the street has the right-of-way. Let him cross.

Watch for pedestrians crossing in the middle of the block. They don't *legally* have the right-of-way. But let them finish crossing anyway.

When driving next to parked cars, watch out for children. Watch out for persons getting out of the parked cars.

Nighttime is very dangerous for pedestrians. Look along the side of the road for people on foot. Slow down at intersections. Look for pedestrians before you drive through. Be careful when turning. There may be a pedestrian around the corner in the street.

Drivers may turn right on a red light after coming to a full stop. Be careful at these intersections. Pedestrians have the right-of-way over a car turning right. Come to a full stop. Let pedestrians cross before you turn.

Make way for ambulances as if they carried someone from your own family.

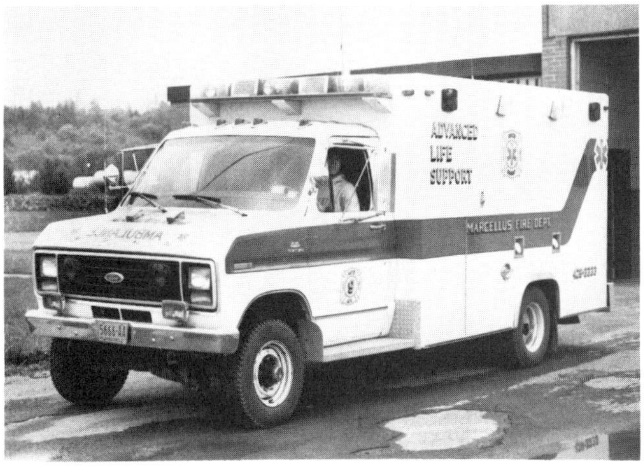

Bikers must follow the same traffic rules cars do. But some bikers don't seem to know that. So watch for them.

Bikes. More and more persons are riding bikes these days.

Bike riders are supposed to obey the same traffic laws that car drivers do. Some bike riders obey only sometimes. You can't predict what a bike rider will do. So you must be very alert. You must be ready for anything. Be ready to slow down or stop when you see a bike rider.

Bike riders have every right to be on the road. So don't try to crowd them off the road. Don't pass too closely. The bike could wobble. The rider could fall.

Funeral procession. A car leading a funeral procession must obey traffic signs and signals. It must wait for the green light or stop at signs. But the rest of the cars in the procession may then pass through the stop sign. They may keep on going even if the light turns red. You must yield the right-of-way to cars in a funeral procession. Cars in the procession have their headlights on.

14
Getting through intersections

Intersections can be very dangerous.

Before entering an intersection, make sure the way is clear. Watch out for other vehicles and pedestrians. Size up the traffic situation before entering an intersection.

Some intersections have traffic lights or signs. They are **controlled intersections**. Other intersections are uncontrolled. They have no stop lights or signs.

Uncontrolled intersections

When you approach an uncontrolled intersection, slow down. Check traffic in both directions. Be ready to stop if you need to.

Obey the right-of-way laws if other cars are also at the intersection. This means:

- Cars moving straight have the right of way over cars turning left.
- A car already in the intersection has the right of way.
- Pedestrians at a crosswalk have the right of way.
- A car on a main street has the right of way over a car on a side street.

Parked cars on the street may block your view of traffic on the cross street. Or another moving car may block your view of the traffic. If your view is blocked, move slowly when you come to an intersection. Be ready to stop or yield the right of way.

If your view is blocked, go slowly through intersection.

Traffic lights

Green means go. Red means stop. Yellow means caution. That's simple enough. So why are there so many accidents at intersections? There must be more to traffic lights than just stop, go, caution.

Yellow. What should you do when the light goes from green to yellow? You certainly can't stop the car in its tracks when the light changes. That's why there is a yellow light. It warns you that the red light is coming. It warns you that you will have to stop. The yellow light gives you time to slow down and stop before the red light comes on.

You should stop for a yellow light if you can stop safely and smoothly. But sometimes, you will have to go through the yellow light. You may be in the intersection when the yellow light comes on. You may be too close to the intersection to stop.

You must use good judgment to decide if you can stop for a yellow light. Let's say you couldn't stop in time unless you slammed on the brakes. If so, the car behind you may not be able to stop before hitting your car.

Some people try to beat the red light. They

Look ahead. If the light has been green a long time, prepare to stop. Don't try to beat the red light.

speed up to get through the yellow light. The driver behind you may think you will beat the light, so he will try to beat it, too.

What if someone is driving close behind you and you plan to stop for the yellow light? Pump your brakes a few times. That will make the brake lights go on and off. That signal will warn the driver behind you that you are stopping.

Plan ahead when you drive. When you are coming to an intersection, look beyond the cars to the light. What color is the light when you are a block away? If the light has just changed to green, you have a good chance of making the light. The light changes every 30 to 60 seconds. If the light has been green for some time, be ready to stop. Take your foot off the gas to slow down. Don't race to make the green light.

Flashing yellow light. A flashing yellow light means you must slow down. Proceed with caution.

Red. If you see from a block away that the light is red, start to slow down. Take your foot off the gas pedal. Slowing down gradually saves wear on the brakes. You will also come to a smoother, safer stop.

When you are stopping for a red light, check behind you. Is a car following too closely behind? If so, tap the brake pedal lightly to warn him with your brake lights.

Right on red. You can turn right on a red light at intersections. You cannot turn on a red light in some large cities. You cannot turn on a red light if a sign says you may not.

When it is legal, you must be in the right-hand lane to make the turn. You must stop at the light first. *You must come to a full stop.* You must yield to pedestrians and other cars. You may not turn until the way is clear.

Flashing red. A flashing red light means you must come to a full stop before going through the intersection. You must yield to pedestrians and other cars. You may go across the intersection only when the way is clear.

A flashing red arrow means cars moving in the direction of the arrow must come to a full stop.

Green. When the light turns green, don't take off too quickly. Glance right and left. Make sure the intersection is clear. When the light turns green for you, it turns red for the drivers on the cross street. But some driver may try to beat the red light. He may slip through the end of the yellow light. He may speed through just as the light changes. Cars waiting to make left turns may still be in the intersection. Wait a moment to allow any cars in the intersection to pass through.

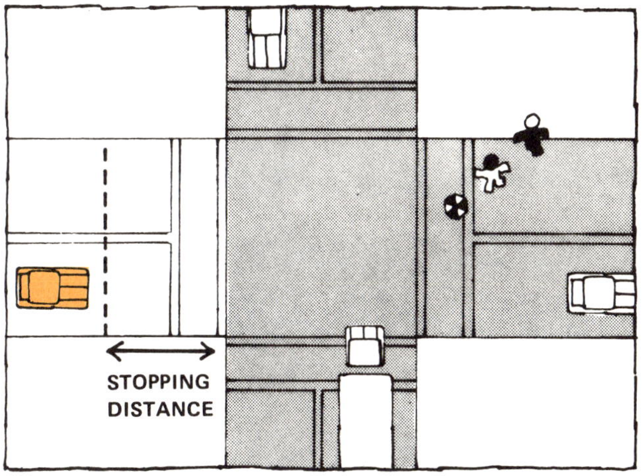

The skillful driver looks way beyond the braking distance and prepares ahead of time for likely dangers.

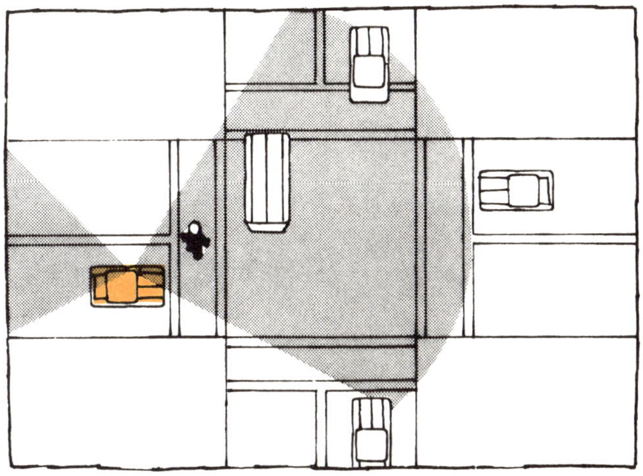

Check in all directions before you wave a pedestrian across. You've stopped, but will others?

Stopping on the crosswalk can cause danger to pedestrians.

Stopping too far out can also be dangerous to you.

Stop signs

Two-way stop signs. At intersections with two-way stop signs, there is a through street. Traffic on the through street does not have to stop. Traffic on the side street, or cross street, must stop.

If you are on the through street, slow down. Look both ways before entering the intersection. A car from the cross street may have entered the intersection.

If you are making a left-hand turn, oncoming traffic has the right of way. You must wait.

If you are on a cross street with a stop sign, you must come to a full stop before the intersection. You must check traffic in all directions. You must yield to cars on the through street.

Come to a full stop at the intersection. If there is a stop line, stop before that line. If there is a crosswalk, stop before the crosswalk. If there are no lines, stop at least 15 feet before the place where the other road crosses your road. Sometimes, when you stop where you're supposed to, you cannot see traffic on the other road. If that happens, inch forward until you can see. Don't go through the intersection until you have checked both directions.

Four-way stops. At corners with four-way stops, cars coming from all four directions must stop. The first car to reach the intersection goes through first. Suppose cars reach the intersection at the same time? Then the drivers follow the right-of-way rules.

A vehicle to your right has the right of way. But there are exceptions to that rule:

– A vehicle already in the intersection has the right of way.
– A vehicle going straight has the right of way over a car turning left.
– Pedestrians in crosswalks have the right of way.
– Emergency vehicles have the right of way.

15 Winter driving

Here's what happens if you let tires spin in one spot.

Millions of us must drive through snow and ice four or five months of the year. There are special skills for driving in snow and ice.

Starting the car

Your car may be hard to start in the winter. Push the gas pedal to the floor once, or twice in some cars. Then turn the key.

If the engine doesn't start, don't let it labor for very long. The engine could be flooded with gas. The car won't start for about 10 or 15 minutes when the engine is flooded.

Starting on a slippery surface

If the road is slippery, you may have trouble getting the car moving. When you try to move forward, the tires may spin.

The key to driving in the winter is driving slowly and smoothly. Don't make sudden moves. Don't brake hard or speed up quickly.

To get moving, push gently on the gas pedal. Keep pressure on the pedal steady. If the car doesn't move, try again. Push even more gently. Try to go even more slowly. Don't let the wheels spin. The road surface will just get more slippery.

If you still can't move, sprinkle some sand on the ice. Sprinkle it behind and in front of the tires that are spinning. The sand (or kitty litter, or gravel) will give the car traction.

Rocking the car

Suppose the car is stuck in snow. If you try to move, the tires only spin. The more you try, the deeper the hole the tires make. The car gets stuck even worse.

Let's say you can't get unstuck by pushing slowly and gently on the gas pedal. Then you will have to rock the car. Rocking the car is moving the

Keep sand or gravel in your trunk to help you get going.

ROCKING YOUR WAY OUT

car forward and backward gently and smoothly. Rocking the car helps to smooth out a track for the car.

Make sure the tires are straight before you start. Shift the car into low. Push gently on the gas pedal. At the moment the car stops moving forward, lift your foot off the gas pedal. Then shift into reverse. At the moment the car stops moving backward, let up on the gas. Then shift into low. Don't quickly shift after spinning the tires fast. This can hurt the transmission. Be gentle on the gas. Keep the car moving back and forth in smooth movements rather than in jerky movements.

As you rock the car, you make a flattened track. Shift back and forth from low to reverse until the car is out of the snow.

Moving on snow

Good snow tires or chains will keep you moving in snow. You must have snow tires to drive on streets marked Snow Emergency Route or Snow Emergency Street. If you get stuck on one of these streets without snow tires, you may get a ticket.

Drive slowly and smoothly on icy or snowy streets. If you try to move too fast, the car will skid. Or, the tires will spin and the car won't go anywhere. If you try to stop too quickly, the car will skid. Turning too suddenly will send the car into a skid, too.

If a road is slippery:

- Drive slowly.
- Don't steer too sharply.
- Don't brake hard. To slow down or stop, pump the brakes. Tap the pedal gently, then let up on the pedal, then tap down on it again, and so on.
- Keep a longer following distance between your car and the car ahead of yours.

Uphill and slippery

Getting up a hill takes extra power. Getting up a slippery hill also takes extra power. But giving the car too much power may keep you from getting up the hill.

As you approach a hill, check it out. Are there any cars stuck on the hill or having trouble getting up the hill? If so, you might wait at the bottom of the hill. That will give the other cars room to back down and try again. It will also allow you to get a running start up the hill.

For extra power getting up the hill, shift into a lower gear. Shift before you start moving up the hill. You might move slowly, but you will be moving. That's what counts.

As you approach a hill, check it out. Sometimes it's better to find another way.

Ice forms more quickly on bridges.

Stopping

You will need more distance for stopping in slippery conditions. So start to brake sooner than you usually would. The safest way to slow down or stop is to pump the brakes.

If you see a car in front of you stopped for a stop sign, stay well behind that car. Give it a chance to get moving. Move slowly and stop near the place where the sign says to stop. Try to stop at a place where traction will be good.

Patches of ice on the streets

The road may look clear, but there may be patches of ice on the road. Those icy patches can be very dangerous. Watch for icy patches. Slow down before you come to an icy patch.

Icy patches can form in shaded areas of the road. Be careful when you drive over or under an overpass or bridge in winter, late fall, and early spring. Bridges and overpasses may freeze sooner than the rest of the road. They may stay frozen longer, too.

Slush

Slush can be very slippery. Drive slowly when the streets are slushy. There could be ice under the slush. Then the road would be even more slippery.

Skidding

To cut down on the chance of skidding:

- Drive slowly.
- Avoid sudden braking or hard braking.
- Avoid sudden turns or lane changes.

If you feel the car skidding, take these steps to control the skid and get control of the car:

- Let up on the gas pedal.
- Steer the car back the way you were headed before you skidded. You want to straighten the car out. You probably would do this without thinking anyway.
- Pump the brakes only after you have straightened the car out and the car is under control again. Don't brake during the skid or you will skid even more.

Trouble conditions 16

There are dangerous conditions in every season.

Rain

Many drivers don't think of rain as a dangerous driving condition. But it can be.

Rain makes the road slippery. A car can skid on a wet road. Cars and trucks drop oil on the road. When there is rain, this oil forms a slippery layer on the road. If you brake hard, turn suddenly, or drive too fast, the car may skid.

A thin film of water can make the car hydroplane. This means the car rides on the film of water instead of on the pavement. If the car is not in touch with the pavement, you don't have much control of the car.

To keep from hydroplaning, keep your speed down. And make sure your tires are in good condition. Hydroplaning can happen at 35 miles per hour even if your tires are in good condition. If the tires are worn or not inflated properly, hydroplaning can happen at even lower speeds.

Rain also cuts down your ability to see. You cannot see as far ahead as you can in clear weather. You cannot see dangers as soon as you would in clear weather. And, the car's stopping distance is longer because the road is slippery.

What does this add up to? It adds up to the need to drive more slowly. It adds up to the need to keep a greater distance between cars.

Use your windshield wipers in the rain. Use the headlights. You will be able to see better. Other drivers and people on foot will be able to see you.

Skidding

Snow, ice, and rain cause skidding. These other conditions also cause skidding: sand and gravel, mud, wet leaves, wet trolley or train tracks, brick or cobblestone roads, and bumps and holes.

You must drive slowly in these conditions. Don't brake hard. Slow down gradually. Avoid bumps and holes if you can. If you must drive over them, drive over them slowly.

Wet roads like these can easily throw a car into a skid.

Why would you drive slowly in these situations?

Fog

If you must drive in fog at night, put the headlights on low beam. Low beam throws the light on the road. If you used high beam, the light would reflect back at you. Use the low beam during the day, so other drivers can see you through the fog.

Drive slowly. Your stopping distance should not be greater than the distance you can see. Let's say you can see only 75 feet ahead of you. Then you should be able to stop within 75 feet. In this case, you would have to drive slower than 30 mph.

If there are patches of fog in the road, slow down before you enter the fog. Watch out for people walking or on bicycles, and for other vehicles. Be sure to stay on your side of the road.

Night

Night is a dangerous time for drivers. It is even more dangerous for drivers with poor night vision.

Sunrise and sunset. You may be looking directly into the sun at sunrise or sunset. The bright sun keeps you from seeing the traffic lights and other cars and people on foot.

Drive slowly. Wear sun glasses. Use the sun visors on the car. Make sure the windshield of your car is clean. Dust and dirt on the glass make it harder for you to see.

What if the sun is at your back? Then keep in mind that drivers in oncoming traffic are looking into the sun. If you must use their lane for passing, make sure you have plenty of time. If you are making a left turn, be sure there is a large enough opening. Oncoming traffic may not see you in time to stop or slow down.

Darkness. In most states, headlights must be on between sunset and sunrise.

When you are in traffic, keep the headlights on low beam. The headlights must be on low beam if you are within 500 feet of an oncoming car. They must be on low beam if you are 200 to 300 feet behind another car.

High beams may be used when there is no traffic close to you. High beams light up the road so that you can see farther ahead.

But high beams shine right into other drivers' eyes. High beams can blind another driver for a moment. So it is very important to dim your headlights.

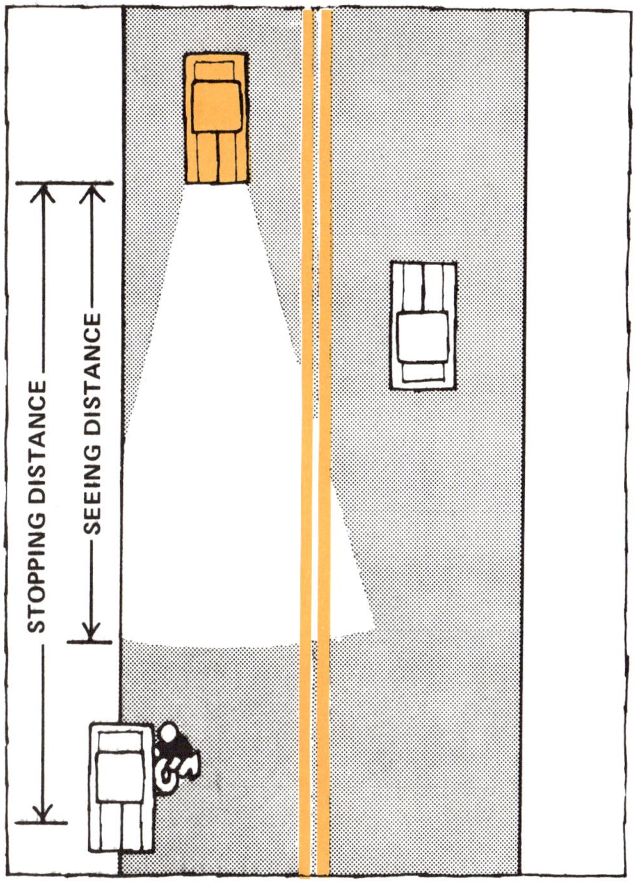

Could this car stop before reaching the hidden person?

(Read about stopping distances on page 62.)

Low beams don't blind the eyes of oncoming drivers.

If a car coming toward you has high beams on, look away from the lights. Look to the white guide line on the right of your lane. Don't look straight into the lights of the other car.

After a car passes by you, you may need a few seconds to adjust to the darkness again. Your eyes must keep adjusting to light and dark at night. Your eyes tire from this. Slow down a little until your eyes have adjusted to the dark again.

Don't put on lights inside your car. They distract other drivers. They make it harder for you to see into the darkness.

Stopping distance. Your car's headlights and taillights help others to see you. The headlights help you to see the road, too. But the headlights do not give you a lot of light. You cannot see as far ahead as you can during the day.

So you must drive more slowly. You must not drive beyond your headlights. This means you should be able to stop within the distance you can see.

Suppose you are driving on an unlighted road. Your high beams light up the road only about 300 feet ahead of you. If you are driving at 55 mph, you can stop in about 210 feet. If you are driving at 70 mph, you can stop in about 345 feet. At 70 mph, you are driving beyond the range of your headlights. (You are also speeding!)

If your headlights are not adjusted properly, you may not see as far. If your brakes are not good, you may need more distance to stop.

If the road is wet or slippery, you will need more stopping distance than on a dry road. On a wet road, your stopping distance may be twice as much as on a dry road. You may need 400 feet to stop if you are driving at 50 mph. Then 50 mph is too fast on an unlighted, slippery road.

There are many factors to take into account when you drive at night. Poor brakes, poor headlights, wet roads, or unlighted streets mean you must drive slowly.

You don't know what is in the darkness beyond your headlights. If you are driving beyond your headlights, by the time you see the danger, it will be too late to stop.

Driving in the mountains

Driving in the mountains means driving up and down a lot of steep grades. You may have to go around dangerous curves, too.

Here are some tips to remember for driving in the mountains or in hilly country.

1. Use the engine to slow the car down. Don't depend on the brakes alone to slow down the car. Before you start down a steep grade, put the car in a lower gear. Then the engine slows down the car. Don't go faster than you should in that gear. Use the brake to keep the speed down to a speed proper for the gear the car is in.

2. Use a lower gear to climb a steep grade. Shift into a low gear for more power in climbing.

3. Don't coast down a steep grade. Coasting means putting the car in neutral and taking your foot off the gas. This is dangerous. It is illegal in many states. The car picks up speed as it moves

Let car ahead gain its downhill speed before you pass.

down the hill. The engine is not helping to slow down the car. Using the brakes at a high speed could damage the brakes. If you shift from neutral into gear, you could damage the transmission. And with the car moving so fast, you may not be able to steer safely.

4. Stay in your own lane. On a mountain road, there are a lot of curves. You can't see beyond the curves. There may be another car coming toward you. Stay in your lane.

Don't pass on a curve or on a steep grade. Pass only when you can see clear road ahead of you. Be sure you have plenty of time. Remember, you need more time to pass on a mountain or steep hill. You cannot speed up as fast.

5. Watch out for objects in the road. You may be driving blind in the mountains. That is, you can't see around the curves or over crests. There may be a rock or animal or car in the road ahead. Drive slowly in case you have to stop. If there is something in the road, go around it. But make sure there is no other vehicle coming toward you.

6. Expect anything. You can run into fog, snow, or ice at high levels. The weather below may be sunny and warm and clear. So be ready for changes in weather and road conditions.

7. Carry a can of water with you. The car has to work extra hard in the mountains. The engine could overheat. If it does, pull way over, off the road. Choose a safe place. Set the parking brake. You will have to wait for the engine to cool down. Then you may have to add some water. Add the water a little at a time. Let the motor run while you add the water.

Desert driving

Good roads cross the deserts now. But there are still problems in driving across the desert. Towns and service stations are scarce.

1. Be sure you start out with a full tank of gas.

2. Make sure the car is in good condition. Check air pressure in the tires. Check the cooling system of the car. Carry a can of water in case the car overheats. Carry an extra fan belt, too.

3. Try to plan your trip so you are not driving across the desert at midday. The sun is at its hottest then.

4. You should probably avoid back roads. You could get lost. You could go off the road and get stuck in the sand.

17 Handling driving emergencies

Handling a driving emergency calls for common sense and staying calm. Don't do anything that will take more control of the car away from you.

Skidding

If the car skids, follow the rules for skidding on snow or ice. The rules are in chapter 15.

Blowout or loss of a wheel

If a tire goes flat or blows out, the car will swerve or wobble. If the wheel falls off, the car will do these same things.

To control the car, follow these steps:

- Hold the steering wheel with two hands.
- Steer straight ahead.
- Slow down gradually by letting up on the gas pedal.
- When you have the steering under control and the car is slowed down, brake gently.
- Drive the car off the street or highway. Get as far away from traffic as possible.
- Put on the four-way flashers to warn others that you are stopped.

Steering failure

If the steering on your car fails:

- Let up on the gas pedal.
- Shift to a lower gear.
- When the car is slowed down, brake gently.

(If you slam on the brakes, the car will swerve sharply. You will have no control at all.)

Brake failure

There are usually two early warnings before brakes fail completely. If you discover one of these problems, have the brakes checked—before they fail completely:

1. The brake pedal sinks slowly to the floor when you push down on it. This is called a spongy brake.
2. You have to push the pedal almost to the floor to stop the car.

If your brakes do fail, take these steps:

- Pump the brake pedal to try to build up some pressure in the brake system.
- If that doesn't work, shift to a lower gear. That will slow you down.
- Apply the emergency (parking) brake. Apply it slowly and gently.

IN CASE OF A BLOWOUT

Check your brakes after going through deep water. They may not work well the first few times you pump them.

Wet brakes

After you drive through a deep puddle, the brakes may not work for awhile because they are wet. Test them after driving through the puddle. Dry them out by pumping the pedal or riding it for a short distance. Riding the brake means keeping light pressure on the pedal as you drive.

Running off the pavement

If you start to fall asleep, or if your mind starts to wander, your car will start to wander—right off the road.

That will wake you up fast!

Don't jam on the brakes or try to turn sharply back onto the road. This is especially true if the shoulder level is lower than the road.

– Hold the steering wheel firmly.
– Take your foot off the gas pedal to slow down.
– Brake gradually.
– Check traffic behind you. Signal and steer back onto the road when it is safe to do so.
– Stop at the next rest area if you are that tired. In the meantime, turn on the radio and open a window for fresh air. That will keep you more awake and alert.

Car coming at you in your lane

If a car is coming at you in your lane, getting angry won't help.

– Pull over to the right as far as you can.
– Honk your horn. At night, flash your lights. The other driver may be asleep.

If your car breaks down

If there is a mechanical problem with the car, get off the road. Get out of the way of traffic.

– Signal and pull off the road onto the road shoulder. (If the engine quits, you still should be able to coast to the side of the road.)
– Once you are off the road, put on the four-way flashers. Open the hood of the car. Or tie a white cloth to the door handle or the radio antenna. These are signs to tell others you're in trouble.
– At night, set up flares. Set the flares a few hundred feet behind the car to give early warning to other drivers.
– Wait for help if you are on the highway. Stay out of the car in case it is struck by another car. Stay away from the traffic side and the front and rear of the car.

Stuck gas pedal

If the gas pedal gets stuck, you might be able to pry it up or jiggle it loose with your toe. But don't waste too much time trying to unstick the pedal.

– Brake gently.
– Shift the car into neutral.

On old cars you can turn the ignition off. Then pull to the side of the road and stop at a safe place.

Newer cars have anti-theft devices. The steering column locks when the ignition is turned off. Then you can't control the steering. So don't turn the engine off. Simply put the car into neutral. Then get the car off the road and stopped as quickly as you can safely do so. Then turn off the engine.

What signals tell you this car is broken down?

Stalled! And a train's coming! What would you do?

Should you always stop for animals?

Headlight failure

If your headlights suddenly fail, try your parking lights or flashing emergency lights. They may give you enough light to guide you off the road. A street that is busy or lighted will probably have enough light to guide you safely off the road.

On a dark and lonely road or highway, slow down. Try to stay on the pavement until you have slowed down enough to move safely onto the shoulder.

Wiper failure or open hood

Suppose the hood pops open. Or maybe the windshield wipers fail in rain or snow You should get off the road quickly. Roll down the window. Stick your head out so you can see ahead of you. Pull off the road.

Stalling on railroad tracks

If your car stalls on railroad tracks, try not to panic. If you hear a train whistle, leave the car. Don't waste time trying to get the car started. Run alongside the track toward the train. Then you won't get hit with pieces of flying car.

Animal in the road

Animals run into the road. Sometimes you will see the animal in time. You will be able to slow down and stop if necessary. Be alert for animals in the road. Be especially alert when you see signs warning of animal crossings.

What if an animal darts into the road and traffic is heavy? If you stop suddenly, you may be rammed in the rear. If you swerve, you may run into another car. You may have to hit the animal.

But usually if you are alert, and moving at a safe speed, you will be able to stop in time.

Sinking car

What if you are in your car and your car is sinking in deep water? Try to get out through a window. If you open a door, water will get in the car faster.

If your car has sunk below the surface before you can get out, go where the air is. There will be an air pocket in whatever part of the car is on top. When the car has settled, you should be able to get out through a window.

Don't open the door. Get out through a window if your car is sinking in water.

Put out a small fire with dirt. But if fire gets near the gas tank, get away fast.

Fire

If steam is coming from under the hood, pull over. The car is overheated. If smoke is coming from under the hood or elsewhere in the car, pull over. The car is on fire.

Turn off the ignition. If you don't have a fire extinguisher, smother the flames with dirt or sand. If the fire is near the gas tank, get away from the car fast.

Don't try to put out a gas fire with water. Burning gas floats on water. The flame can easily spread.

18 Accidents happen

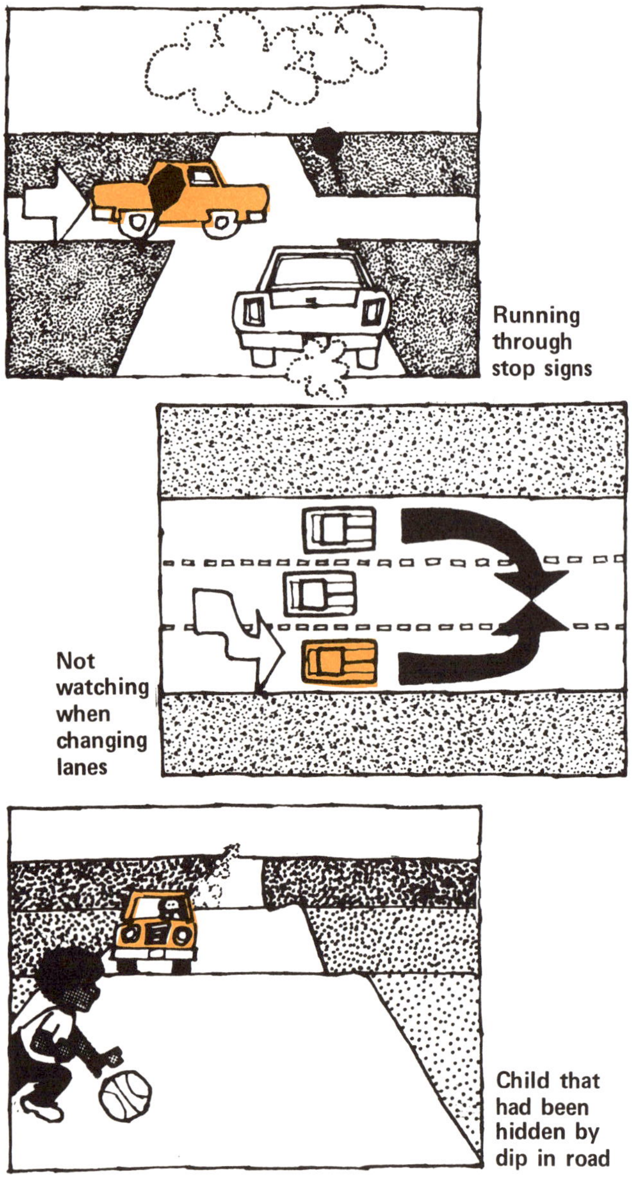

Thousands of surprises can lead to accidents. How could accidents be prevented in the surprises above?

Accidents happen. But they always happen to someone else. At least, that's what we would like to believe.

But what if you are involved in an accident? What should you do?

It can happen to the best drivers

Most accidents can be prevented. Accidents happen when a driver is driving too fast or is not paying attention. Accidents happen because of errors in judgment. Sometimes accidents happen when something goes wrong with the car.

If you drive defensively, you stand a good chance of avoiding or preventing an accident.

But, still, even the best of drivers have accidents. A child may run into the street very suddenly. An out-of-control car may speed through an intersection. Or a good driver may just make a mistake.

Suppose an accident is about to happen, and you can't avoid it. You must make some quick decisions. You need to think quickly, clearly, and logically. You must maintain control of your car and act quickly.

Force of impact

You must choose the action that will cause the least injury or damage. The seriousness of the injury or damage depends on the force of impact. The force of impact is the force with which two objects collide.

The force of impact depends on:

– speed of the car, or the speed of both objects as they collide with one another

– the weight of the car and other object

Damage done to car during crash at different speeds

– the give of the object that the car hits

– the angle at which the car collides with an object or another car.

You can reduce the force of impact by taking these steps:

1. Slow down. The force of impact is much greater at 40 mph than it is at 20 mph. Slow down as much as you can before hitting anything.

But you must make a decision. Should you slam on the brakes? Slamming on the brakes will lock the wheels. You will not be able to steer the car well. You will probably skid. You may skid into the other lane or into an object. But you may be able to stop without hitting anything.

Perhaps you should slow down as much as possible without locking the wheels. Then you will still have steering control. You could still guide the car off the road and into a bush or snow bank.

You must decide. Your decision depends on what the traffic and the road is like. Which move will cause the least injury–hard braking or good steering control?

2. Hit an object that has some give rather than a solid object. Try to hit a bush or a snow bank rather than a tree or pole. An object with give will slow you down and then stop you. A solid object will stop you short. It stops you hard.

Try hitting the wall with your fist. Then try hitting the bed or a pillow. It's not hard to decide which punch hurt the most!

If you are going to steer off the road to avoid hitting a person or a car, steer to the right. If you steered to the left, you would be in the way of oncoming traffic. You might get hit, unless you are on a road with a grass island or bushes between you and oncoming traffic. In that case, you could steer to the left.

3. Try to hit with a glancing blow rather than head-on. Hitting a tree or another car head-on is the worst kind of crash. You probably wouldn't be able to walk away from such a crash. A glancing blow gives you a better chance of staying alive.

To hit with a glancing blow means to hit at an angle. It means to hit the side of the object with the side of your car. It means not hitting the object square. If you hit at an angle, the force of impact will not be as great.

Pedestrians

Pedestrians and bike riders are not protected. If they were struck by a car or truck, they would not stand a very good chance. The force of impact would be great, especially if the car was moving fast.

When you are driving a car, you are protected by that car. You must be alert to the person on foot or on a bike who is not protected. Be even more alert at night and in the city. Be keenly alert in areas where children are playing. If you see children in yards or on the sidewalks, slow down. Be ready to stop. A child may dart in front of your car.

Suppose you can't stop before hitting a person on foot or on a bike? Then steer with a glancing blow into a parked car or another object. You are

Split second decision: Which way would you go?

likely to escape being hurt if you hit a parked car. You're protected by your car. But there's little chance for the walker or bike rider if you hit him. He has nothing to protect him.

In case of an accident . . .

If you get in an accident, you should take certain steps:

1. You must stop. Hit-and-run is a serious crime.
2. Find out if anyone has been hurt. Attend to any injured persons. Do not give first aid unless you are well-trained to do so. Do not move an injured person unless there will be further danger if he stays where he is. Cover the injured person and keep him quiet.
3. Call the police and an ambulance if needed. The police must be called if anyone is injured or if property damage reaches a certain amount. That amount varies from state to state.
4. Set up flares or use flashlights to warn others to steer clear of the accident.
5. Have someone direct traffic until police arrive.

Get this information

Drivers involved in accidents should give each other certain basic facts. Get this information:

– names and addresses
– license numbers
– license plate or registration numbers
– year, make, and model of cars involved
– names of insurance companies
– names and addresses of any riders in the cars.

Ask witnesses for their names and addresses. If they will not give you this, write down the license plate numbers of their cars. Then you can trace these witnesses later if you need them.

Write down the name and badge number of the police officer who checks out the accident.

Note details about the accident:

– the time of day
– weather and road conditions
– direction you were driving and the direction of the other car or cars. (Make a quick sketch of what happened.)

What to say and leave unsaid

At the scene of the accident, do not admit guilt. Do not accuse the other driver. Do not say how much insurance coverage you have.

Report the accident to your insurance company within 24 hours. If you are away from home, report it to the company's nearest office.

Use the diagram below to show the location of your vehicle at the time of the accident. Also show the location of any other vehicles, pedestrians, or obstacles involved in the accident. Show any traffic control devices (traffic lights, stop signs, etc.). Indicate the direction of vehicles thus:

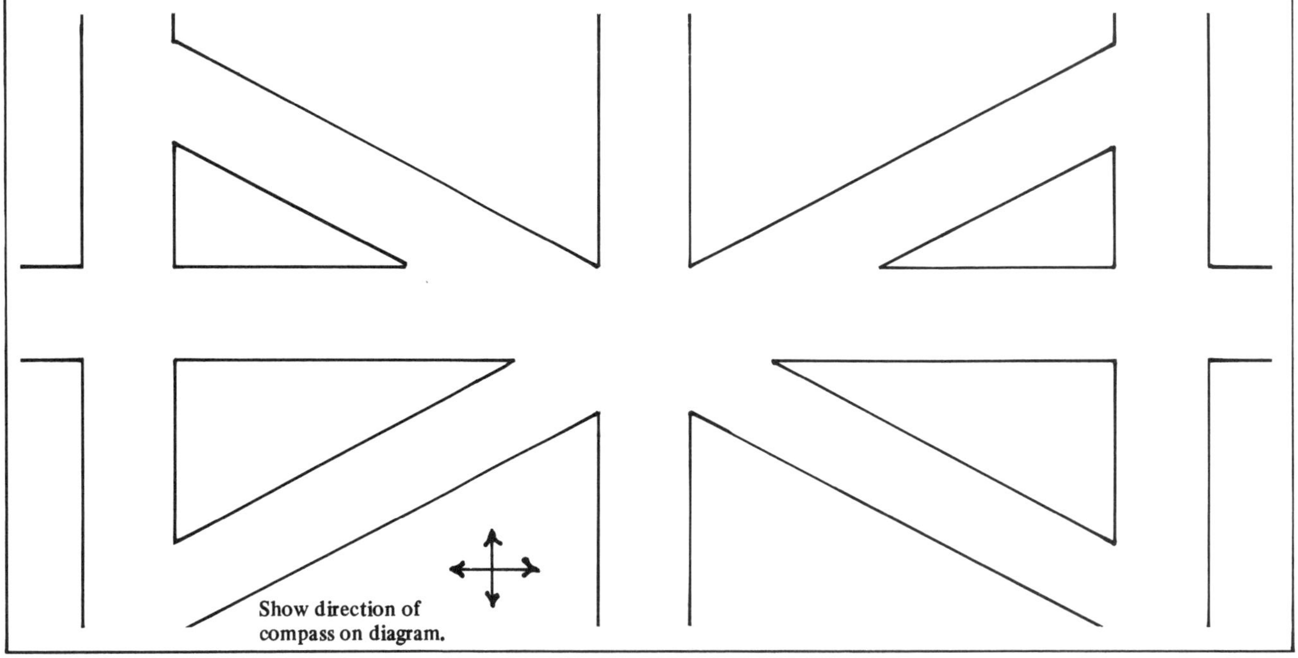

This part of the accident report lets you show how it happened.

If police are there, don't stop unless they tell you to.

You must file an accident report with the police or the department of motor vehicles:

– if someone is injured or killed, or
– if property damage is greater than an amount set by the state. That amount can be from $25 to $200, depending on state law.

You can get an accident report form from the police, the department of motor vehicles, or your insurance agent. The time limit for filing the report varies from state to state. It can be from 24 hours to 30 days. The report will ask for much of the information mentioned before. Your insurance agent or a lawyer can help you fill out the report.

If you bump a parked car, try to find the owner. What if you cannot find the owner? Then write a note which includes your name, address, and telephone number. Place the note on the car you hit.

If you witness an accident

If you see an accident happen, stop and help. Get your car out of the way. Give help to the injured if needed. But give help only if you are qualified. Offer to call the police or to call an ambulance or to direct traffic. If you must call the police, note exactly where the accident happened. Stay on the scene until help comes and until the police say you can leave. Cooperate with the police and the persons in the accident. Give your name and address.

At the scene of an accident

If you did not witness the accident, but you arrive there first, offer help as already noted.

If you arrive after the police have arrived, keep on moving. Don't stop to gape. Just keep moving so that traffic doesn't get tied up. Stop only if you are directed to do so.

Covering accident costs

All states have financial responsibility laws. Each state sets an amount that a driver must be able to pay if he is in an accident. This amount of money goes for damages and injuries to others and their property. The amount can be from $5,000 to $25,000.

Liability insurance is proof that you can pay for damages. In some states, you must have insurance in order to register your car. The state sets the minimum you must have.

In all states, after an accident, you must prove that you can pay the minimum set by the state. Most drivers have insurance. That is their proof. If you don't have insurance, you must deposit money with the state. If you don't have the money, you lose your license.

You must provide the proof soon after the accident. You must provide proof even if it is not yet decided who was at fault.

Proof of financial responsibility must be given:

– if anyone was killed
– if there were any injuries
– if property damage reached an amount set by the state.

A written report must be given to the insurance company soon after the accident. You might not think the accident was serious. But a claim might be filed against you later. Or, weeks later, you may find that you were injured. The injury just did not show up right away. What if you did not report the accident soon after it happened? Then the insurance company might refuse to help you or pay you.

Your insurance agent will help you file a claim. This means he will tell you how to ask for money from your insurance company. Or he will help you get the money from the other driver's insurance company if the other company must pay.

Hidden injuries

If you were involved in an accident, it is good to see a doctor. There might be no sign of an injury at first. In fact, an injury might not show up for many days or weeks after the accident. If you received a hard jolt or bump on the head, there might be hidden injuries. To be safe, see a doctor. Seeing a doctor right away will help protect any insurance claims for injuries. Later, if you discover you were hurt, you will collect your money more easily.

Defensive driving

Defensive driving–we haven't used these exact words. But we keep coming back to that idea. It is an important idea in driving. You must watch out for others when you drive. A car is a heavy piece of machinery. It can be dangerous to you and to others.

You make mistakes when you drive. Other drivers make mistakes. Sometimes drivers are careless or reckless or not very skilled. You have to watch out for them. You have to be prepared for anything when you drive. You must be alert in order to avoid accidents.

Check out the traffic situation as you drive. Look ahead. Size up the situation. If there are children, expect a child to run in front of you. If you are driving up a hill, expect danger over the top of the hill. If you are at an intersection, expect someone to go through the red light.

If you see a car backing out of a driveway, be ready to stop. That car may back out in front of you. If a car a few cars ahead of yours slows down, be ready to slow down, too. If you see a car with an out-of-state license plate, be very alert. That driver may not know his way around town. He may slow down or turn without warning. Take in the whole traffic scene when you drive.

If you must slam on your brakes often, you probably drive too fast most of the time. Or you are not very alert. A driver may have to stop suddenly now and then. But an alert driver keeps an eye out for possible dangers. He doesn't have to brake hard very often.

Most accidents can be avoided if the driver is alert. The sooner a driver spots danger, the less chance there is that a danger will turn into an emergency.

Part Three

You've heard it said that the only thing wrong with some cars is the loose nut behind the wheel. That's too often true. And it's not funny. Too many of those nuts get cracked. Safe driving takes a **sound mind**. It demands a fairly **healthy body** and **solid control over your feelings**. Want to see if you have what it takes? Then here's . . .

Your Driving Fitness Checkup

Sick and tired? Watch out! 19

Would you feel safe in this driver's hands?

When you first start to drive, just keeping the car on the road takes almost all you've got. Remembering the traffic rules is pretty hard.

But after you have some practice, driving becomes easier. The rules make sense. You don't break into a cold sweat at the thought of turning left at a busy corner.

To be a good driver, you need confidence in your handling of the car. You need to know the traffic rules. You need to follow the traffic rules. But other factors affect your driving, too. Those factors are your physical fitness and your mental attitude.

This part of the book helps you check out your body's condition and your mental attitude. You'll see how your body, your thoughts, and your feelings can affect your driving.

Your health

What happens to you when you have a cold?

Maybe you feel tired. Your nose and head may be stuffed-up. You may feel a little grouchy. You are not as alert as you are when you are healthy. And you keep thinking about how awful you feel.

When you are sick, you may not watch the road as carefully as you should. You may not drive as well as you can when you are healthy.

If you are not feeling well, you must decide if you feel well enough to drive. You might decide to stay home in bed. Or you might decide to ride with someone else.

If you decide to drive, you must make a special effort to keep alert. But it's best not to drive long distances. This will only make you more tired and less alert.

Sleepiness

A tired driver can be a very dangerous driver. He must devote much of his energy just to staying awake.

When a person is tired, he cannot see as clearly as he usually does. His reflexes are not as fast. His judgment is not as sharp. His mind wanders from the road.

What if the driver loses his battle to stay awake? Then no one is controlling that one- or two-ton vehicle. That means trouble.

Keeping in shape for alert driving

Don't start out on a trip when you are tired. You will only get more tired.

Suppose you start to get tired after you have been driving for a few hours. What should you do?

If you are on a highway, stop at a restaurant or rest stop. Get something to eat. Or just get out of the car and stretch your legs. Or take a short nap.

What if there is no rest stop or restaurant close by? It's safer to stop by the side of the road than to keep driving.

But be sure that you pull way off the road. Don't stop on a curve or hill. That makes it hard for other drivers to see you in time to avoid hitting you.

Stop where other drivers can see you from a far distance. Put on your parking lights to warn other drivers. And be sure to lock your doors.

If you feel yourself getting tired, the best cure is sleep. But if you can't stop and rest, here are some steps to take:

– Listen to the radio. Change the station often.
– Open your window for fresh air.
– Chew gum and eat snacks.

Doing these things will help. But they can't take the place of what you really need–sleep.

Highway hypnosis

If you drive for many hours, you may become hypnotized by the road. Your eyes will be fixed in a blank stare. You'll find yourself driving by habit. But your mind won't be on what you're doing. If you are tired, there is even more chance that you'll be hypnotized.

Staring at the lane lines may hypnotize you. Or the motion of the windshield wipers may hypnotize you. You watch the wipers or the lane instead of watching the traffic.

You may wander from your own lane. You may take your foot off the gas pedal and slow down suddenly. The car is not under your control when you are hypnotized.

To fight highway hypnosis, follow the tips to fight sleep. And try these ideas, too. Change the position of your hands on the wheel. Change the position of the seat or your sitting position. But don't change into unsafe hand or sitting positions.

Lane markings or wipers can hypnotize you.

With a specially equipped car, this person can drive as well as anyone else.

Handicaps and disease

People without the use of their legs can still drive a car. People without the use of arms can also drive. They can drive specially-equipped cars.

People with certain long-term diseases can safely drive, too, if they are under a doctor's care. The doctor's treatment allows them to function normally.

Age

Most persons' physical condition changes as they get older. They may be in good health. But their eyesight gets worse. Or their body movements slow down. They may tire more easily than they used to.

Their hearing may get worse. Their night vision might get worse.

We don't mean to paint a bleak picture. But, our ability to drive safely may change as we get on in years. If so, we need to learn how to make up for these changes. The next chapter shows several ways of coping with these changes.

20

How's your seeing and hearing?

You don't have to see or hear perfectly to be a good driver. But all drivers should understand any hearing or seeing problem they may have. This will help them adjust their driving habits to make up for their problem.

What makes for good vision?

There are many factors that make up your vision. We will look at each of them.

Sharp vision

How well can you see an object in front of you? Do you see it sharply and clearly? Or is it blurry?

In every state, you must pass a vision test to get a driver's license. What if your vision is not sharp enough without glasses? Then you must wear glasses or contact lenses when you drive.

Do you think you need glasses? Visit an eye doctor before you try to get a license. Then you won't need to worry about failing the vision test.

Field of vision

Field of vision is the entire area you can see at one time. It is the area you can see to the front and sides while looking straight ahead.

Have you heard the expression "out of the corner of your eye"? That refers to your side vision.

Some persons can see a lot to either side of them. Some persons can see very little. If you can only see in front of you, you have **tunnel vision**.

Side vision helps alert you to possible dangers.

For some people, eyeglasses make a big difference.

The ability to see to the sides is important to safe driving. It allows you to see cars entering the street from side streets or driveways. It allows you to see children on the sidewalk.

With tunnel vision, it's hard to see people or things heading toward you from either side. If your side vision is good, you have more time to react. You can avoid trouble coming at you from the sides.

But a person with poor side vision can still be a very good driver. He must make up for his poor side vision. He must keep moving his head from side to side to see children, animals, or other vehicles.

Depth perception

Depth perception is ability to judge the distance between you and something else. A driver must be able to judge his distance from the car in front. He must be able to judge whether he has enough room to pass another vehicle. He must be able to judge whether he can turn left before an oncoming vehicle reaches him.

If your depth perception is poor, you must make up for it when you drive. Allow extra room for braking, backing up, passing, and turning. Stay farther away from the car in front of you than you think you should. The usual distance is one car length for every 10 miles per hour of speed. Stay farther away than that. Suppose you are not sure you have time to make a left-hand turn. Wait for the other car or cars to go by, then turn.

Color perception

Maybe you know someone who is color-blind. Maybe you are color-blind. A person who is color-blind doesn't see colors as others do. A common form of color-blindness is red-and-green color-blindness. A person with this vision problem can't recognize red or green.

You have to take a test for color-blindness when you apply for a driver's license. If you are color-blind, you can still get a license. But you must make up for the problem.

Colors give us messages when we drive. Different-colored traffic lights tell us when to stop and go. Signs give us instructions. Sometimes their colors tell us what the sign says before we can make out the letters.

A person who is color-blind must learn the order of traffic lights instead of watching the color. Red is always on the top. Yellow is always in the middle. Green is always on the bottom. He must watch which is lit up.

A simple side vision test. Most people can see the pencil at point A while looking straight ahead.

With good depth perception you could see how far B is from C. But with poor depth perception cars B and C might seem to be closer together than they really are.

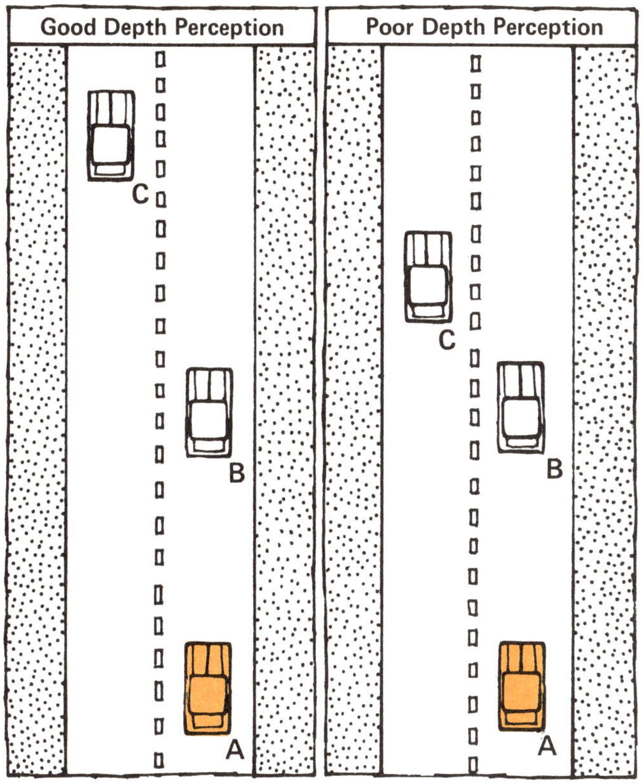

What about signs? Different kinds of signs have different shapes. Warning signs are diamond-shaped. Stop signs are eight-sided. A color-blind driver can get a hint about what a sign says by its shape.

Night vision

You can't see as well at night as you can during the day. Even on well-lit streets, there is not as much light as there is during the day. Some streets are not lit up at all. Your headlights provide the only light showing you where to drive.

But there is even more to the problem. Lighting changes quickly when you drive at night. You may turn off a well-lit street and onto a dark one. It takes a few seconds for your eyes to adjust to the new lighting. Or a car may be coming toward you with its high beams on. They flash into your eyes. When the car has passed, you must once again get used to the darkness.

Rain and wet roads make the vision problem even worse. The headlights reflect off the wet pavement. This can cause glare. Poorly adjusted low beams of oncoming cars can shine in your eyes and blind you for a second.

Night driving is hard, even for persons with good eyesight. It is even worse for persons who don't see well in daylight either. Many persons have a very hard time seeing at night.

Learn how good your night vision is. If you don't see well at night, try not to drive at night. The darkness and the glare keep you from seeing well enough to drive safely.

Your eye doctor might have a machine that tests your night vision. These machines test your vision in poor lighting, your glare vision, and how quickly you recover from glare.

States do not test for night vision. But there are many people who should not drive at night. You must decide for yourself if you are one of them.

Tired eyes

After many hours of driving, or after driving in tough conditions, your eyes can become very tired. You are tired. Your reflexes are not very fast. This is when you need sharp vision—to warn you early of danger. But when you are tired, your eyes start to fail you, too.

Your headlights may not show all that's there.

Vision changes. Eye checkups are important.

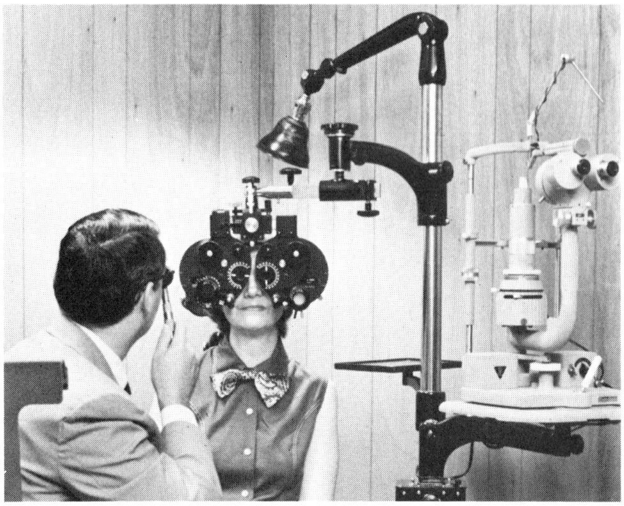

When your eyes are tired, your vision, your depth perception, and your field of vision get worse. And when your eyes are tired, you are in danger of falling asleep at the wheel. You are tempted to just close your eyes and rest them. If that happens, pull over for a while and rest.

How well do you see?

Many people are not aware of their vision problems. Their vision may be a little blurry. But they may have become used to it. Maybe they have never seen clearly. They have no way of comparing their vision to perfect vision.

It is important to be aware of any vision problems. Then you can make up for them in the ways we have talked about.

Remember, also, that our vision keeps changing. It usually gets worse as we get older. Field of vision may get narrower. And depth perception and night vision worsen. So it is a good idea to visit an eye doctor every year. Your eyesight may worsen—or get better. You may need glasses or you may need to have your glasses changed.

You may be amazed by how different the world outside your windshield looks when you can see clearly.

Driving and hearing

Have you ever stopped to think how much we rely on our hearing when we drive?

Our ears allow us to hear sirens. Those sirens warn us of approaching ambulances or fire trucks, or police cars. We can hear a warning honk from another car. Our ears tell us if our car is not running well. We also hear brakes squealing or children playing near the road.

But deafness or poor hearing is no reason for being a poor driver. A person with a hearing problem can learn to make up for this problem.

The person with a hearing problem must rely on his sight. He must rely on his side vision. He must turn his head more to look for rescue vehicles, other cars, children, and animals. A deaf person must use mirrors more. He may even have extra mirrors put on his car. Other drivers can hear if the engine is acting up or the fan belt is slipping. But he must watch the oil gauges and other gauges to make sure the car is running safely.

Drivers with hearing problems have been found to drive as safely as persons with good hearing. Perhaps the deaf driver knows he must be extra-alert. And he is not distracted by other noises such as the radio. He also makes better use of his sight in many cases.

Like vision, hearing may worsen over the years. So it's best to keep aware of changes in your hearing and adjust to them. This will make you more able to drive safely.

Would good hearing help in these situations?

Dare you drink and drive? 21

"If you drink, don't drive." Ads on television and in the newspapers say this. And there are good reasons for this warning. Alcohol affects your physical condition. Alcohol affects your attitude.

Physical reactions

Drinking can cause changes in your body that affect your driving.

Vision. Drinking can do strange things to your vision. Drinking can affect your depth perception. You may not be able to judge how far away a person or car is.

Your ability to see to either side may be affected, too. Your field of vision can become narrower after you drink. Then you have trouble seeing people or cars until they are in front of you. You have less time to react to things coming from either side.

Drinking can make you see double. It can make your vision blurry. You will have trouble seeing signs, other cars, pedestrians.

Most people do their drinking at night. It is hard to see at night when you are sober. If you have been drinking, it is even harder to see at night.

Reaction time. Drinking slows down your reactions. You need more time to realize the light is red and you must stop. You may need extra time to get your foot to push the brake pedal. You need more time to react to a person walking on the street. You do not respond as quickly to traffic dangers.

You really need to be able to spot problems sooner if your reactions are slowed down. But your vision is poor, too. You can't see problems as soon. So you are robbed of crucial seconds by your poor vision and your slow reactions.

Drinking also affects your body movements. You may have trouble doing simple tasks. You may have trouble turning on the signal light. Your foot may slip off the gas pedal or brake pedal. You may hit the gas pedal instead of the brake pedal.

Attitude and moods

Drinking affects your attitude and your judgment.

Attitude. After only one or two drinks, you may not think as clearly as you usually do. But you may think that your driving skills are better than ever. So you may end up taking more chances.

You may drive too fast. You may pass when it isn't safe to pass.

Moods. Drinking makes some people happy. It makes other people angry or depressed. It may make a person happy one time and depressed at another time.

These moods affect our driving when we are sober. After a few drinks, our moods may be extra strong. And it is hard to put our moods aside so that we can drive.

How much is too much?

Many persons brag that they can drink all night and not get drunk. But all that drinking does have an effect. It takes very little alcohol to make it dangerous for a person to drive.

Unlike food, alcohol does not have to be digested. Alcohol enters the bloodstream right away. Your body carries the alcohol until the alcohol is burned up. If you drink too much, the alcohol builds up in your body. It produces the effects we just talked about.

If you drink slowly enough, you may not be affected by the alcohol. For most persons, the body burns up only about one-half ounce of pure alcohol an hour. That is the amount of pure alcohol in one mixed drink. If you drink one mixed drink an hour, you will probably not get drunk. If you drink several beers or mixed drinks, the alcohol builds up. You are on your way to getting drunk.

You can control the rate at which alcohol enters your bloodstream. The rate depends on many factors.

Person's size. A large person is less affected than a small person by the same amount of alcohol. The larger person has more blood. The alcohol is diluted more by his or her blood.

Have you eaten? Alcohol enters the bloodstream more slowly through a full stomach than through an empty one.

Kind of drink. Beer and ale and wine enter the blood more slowly than the "hard stuff."

Mixed drinks enter the blood more slowly than unmixed drinks. Alcohol mixed with water enters the blood more slowly than alcohol mixed with soda.

Speed of drinking. If you sip your drink, you can make the drink last longer. If you drink slowly, your body has time to burn it off. If you gulp the drink, your body cannot burn off the alcohol fast enough. The alcohol builds up in the body.

Estimated blood-alcohol concentration

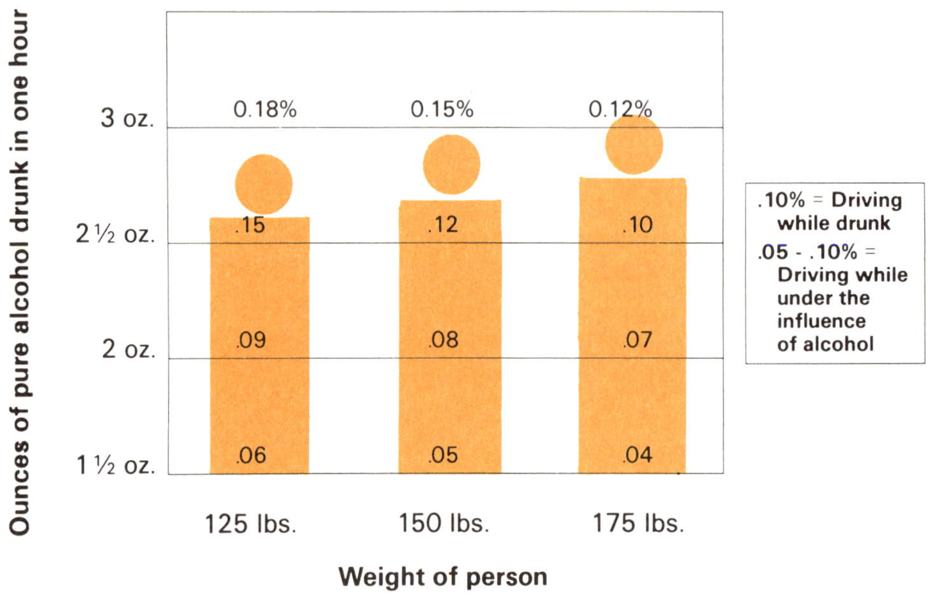

1 ½ ounces of pure alcohol is contained in: 3 cans of beer *or* 3 glasses of wine (3 oz. each) *or* 3 shots of whiskey (1 oz. each).

2 ½ ounces of pure alcohol is contained in: 5 cans of beer *or* 5 glasses of wine (3 oz. each) *or* 10 shots of whiskey (1 oz. each).

A few drinks can affect even the best of drivers. After a cocktail party this driver couldn't steer around boxes on the course that he had easily avoided before drinking.

Legally drunk

You may have a drink or two in an hour and not think you have had too much. But you may have had enough to be legally drunk.

The law decides if you are drunk by measuring the concentration of alcohol in your blood. The police use a breath test. This test finds out the blood-alcohol concentration (BAC).

A person is probably too drunk to drive if his blood-alcohol concentration is 0.10. Suppose you weighed 150 pounds. If you drank two martinis, or five bottles of beer in an hour, your BAC would be 0.10%.

Most states say that a BAC of 0.10 means a person is legally drunk. He could be arrested for driving while drunk. Some states use a higher or a lower BAC.

Even if the BAC is below 0.10, you may still be too drunk to drive safely. If the BAC is between 0.05 and 0.10, you could be arrested for driving under the effects of alcohol.

Law enforcement

Drunk drivers are a danger to themselves and to other drivers and pedestrians. Drunk drivers are involved in about half the fatal accidents on the roads each year.

Police and the courts are cracking down on drunk drivers. In many places, a person who is arrested for drunk driving loses his license for six months to a year. His insurance rates will go up. And the drunk driving arrest will be on his driving record.

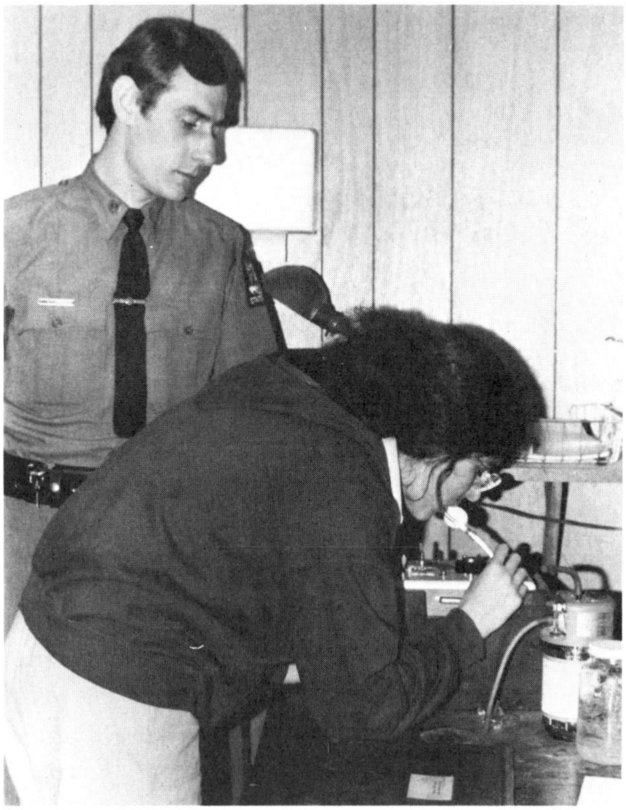

This machine measures the amount of alcohol in a person's breath. This tells how much alcohol is in the body.

22 How about drugs and driving?

Mixing drugs and driving can be dangerous. By drugs we mean:

— over-the-counter drugs
— drugs taken under a doctor's orders
— drugs that are misused or illegally used.

Drugs can affect your physical condition, your attitude and your emotions.

Different drugs cause different effects. But some common effects are sleepiness, seeing unreal things, and loss of control over your body. Drugs also cause shaking, over-excitement, panic, and strange fears.

Taking illegal drugs, or drugs that are not ordered by a doctor, can be dangerous. You can have unexpected reactions. You can lose control of your body and your mind. There are state laws against driving while under the control of drugs.

Over-the-counter drugs

Over-the-counter drugs are drugs you can get without a doctor's order. You can buy them in drugstores or supermarkets. They are cough medicines, aspirin, laxatives, and so on.

Read the labels on any over-the-counter drugs that you buy. They do cause side effects. The side effects could make you less able to drive safely.

Many over-the-counter drugs can make you sleepy. Even plain aspirin makes some people sleepy. Some over-the-counter drugs may make you feel shaky.

Some over-the-counter drugs can affect your driving.

You need to find out how you react to any drug that you are taking. Don't drive until you are sure the drug does not cause bad effects. If the drug does cause bad effects, don't drive while you are using it.

Maybe you are taking a drug that your doctor told you to take. Ask the doctor or a druggist about the drug's side effects. Is it safe for you to drive after you have taken the medicine?

Some drugs can make you sleepy. Some blur your vision. Any number of effects can happen after taking medicine your doctor ordered. Be sure it is safe for you to drive after you take the medicine.

Misused drugs

Some persons take drugs for fun to get high. Some persons use drugs to help them sleep. Some persons use drugs to help them stay awake. Drugs are taken for many reasons. And they are very often misused.

Misusing drugs can be dangerous. Misusing drugs and then driving means double trouble.

Amphetamines. Amphetamines are also called **pep pills**. Sometimes doctors order them for treating mild depression or for helping dieters. But some people use pep pills illegally. Taking pep pills can cause weird pictures in your mind or jitters or highs.

Drugs may make the world look more exciting. But who wants to drive in such a world?

Pep pills can affect your driving. A small problem can become a great one because you may overreact. You are too jittery to remain calm. You can't handle the problem. You are too excited to make a decision about what to do. You are too shaky to react.

Barbiturates. Doctors order barbiturates to help persons sleep or relax. We have already talked about the dangers of driving when you are sleepy.

Hard drugs. Hard drugs are morphine, heroin, and cocaine. Heroin is not legal for any use in this country. Morphine and cocaine are legal drugs in hospitals. Today, cocaine is widely used illegally.

People who misuse these drugs can be very dangerous drivers. Cocaine affects the nervous system. Its effects are similar to, but worse than, those of pep pills. Heroin and morphine have even more drastic effects.

Marijuana. Marijuana, like alcohol, can cause different reactions in different people. Some people think that smoking marijuana makes them better drivers. Others think that they drive poorly under the effects of marijuana. Some users fall asleep or get very drowsy. Some get more lively.

A person cannot predict how he will react to marijuana. And his reaction may differ from one time to the next.

A user may think that he drives better after smoking. He may think he drives more slowly and carefully. But he doesn't know for sure. Mixing marijuana and driving is not a good idea.

LSD. LSD causes strange visions of an unreal world, or "trips." Driving under the influence of LSD is a definite NO.

Non-drugs, but problems anyway

Coffee and cigarettes aren't really drugs, but they can still cause problems to the driver.

Coffee. Coffee can help you if you are tired. It can keep you awake. But too much coffee can hurt your driving as much as sleepiness can.

Too much coffee, especially on an empty stomach, can make you nervous or jittery. It might make your hands shake. It might make your stomach sour. It might even make your eyes play tricks.

Cigarettes. Smoking can have a slight effect on your physical ability to drive. It slows down your body a little. But the real problem is that smoking takes your mind off driving. Lighting a cigarette, dragging on it, putting it out—all these acts can distract you. Smoke can get in your eyes and blind you for a moment.

Suppose you drop the cigarette. Then you try to look for it while the car is going. That will really distract you from the task of driving. Smoking while driving can cause problems.

Drinking, drugs, and driving

We have seen that mixing drinking and driving is dangerous. And we have seen that mixing drugs and driving is dangerous. Put drinking, drugs, and driving together, and you are really in trouble.

There is no telling what effect mixing drugs with a drink can have. It could knock you out. It could kill you. Mixing drugs and drinking with driving could knock you off the road forever.

23

Defensive driving: the safest way

To become a safe driver, become a defensive driver. That means watch out for other drivers, pedestrians, and bikers. Be ready for their mistakes. Mistakes cause accidents. But driving defensively can prevent accidents.

Defensive drivers avoid danger

Defensive driving starts with you. Here are some ideas for keeping yourself from causing danger.

Stay alert. Don't drive if you are too tired. If you have a problem, try not to think about it when you drive. Keep your mind on the task of driving.

Avoid distractions. This is part of staying alert. Don't get too involved in the music on the radio. Don't get too involved in talking with your riders.

Watch out for the unexpected.

Avoid distractions.

If you are driving with children, make sure they are wearing safety belts or sitting in safety seats. This keeps them from jumping around a lot. Don't let them wave their arms in front of you. And try to keep peace in the car. If there is a fight, pull over and stop before you try to break up the fight.

Know yourself and your car. Don't push yourself too hard. Don't fool yourself into thinking you have more driving skill than you really have. Learn to recognize when you have driven long enough. If you have hearing or seeing problems, learn what your limits are.

And know what your car can do. If your car doesn't have a lot of pick-up, don't pull onto a main street until you are sure you have enough time. If you must make a U-turn, know if your car can do it in one move.

Should you slow down?

If you didn't slow down, it's too late!

Watching out for the other guy

Here are some tips on how to avoid dangers caused by others or by poor road conditions.

1. **Look out for possible dangers.** Look as far as you can see down the road so you can spot possible dangers. Look around you—to the side and the rear, too. Try to see as much as you can of the driving situation. Keep far enough behind buses or trucks so that they don't block your view.

Watch for these situations: Are there children playing in yards or on the sidewalk? Are there bikers in the road? Is there a car waiting to pull onto the street from a side street or driveway? Is there a dog in a yard?

Be ready for a child to run out in front of your car. Watch for the biker to wander into your path. Be ready for that car to pull out in front of you.

2. **Don't assume.** You are waiting to turn left onto a main street. A car coming toward you has on its left signal light. But don't assume that it is safe for you to turn. The other driver may not be turning. He or she may not even know the signal light is on. Wait until you are sure the other car is turning and not going straight.

Just because you have the right of way at a corner, don't assume the other driver will let you go first. Wait and see. Don't assume that the other driver will stop right away for a red light. Some drivers try to beat the light and zoom on through.

3. **Adjust to conditions.** If the road is slippery, or the night very dark, slow down. Don't drive too fast for the road or weather conditions.

4. **Don't sneak up.** Make sure others know you are there. If it is raining or snowing, you might turn on the headlights. At night, turn on the headlights. Flash the lights if you think someone doesn't see you.

During the day, or if the weather is clear, you might tap the horn to let others know you are coming. Let pedestrians, bikers, and other cars know you are coming.

The main rule

And the main rule of defensive driving? Watch out for others. You never know when they may do something dumb!

Let the person ahead know that you are passing.

Notes

Notes

Index

—E—

—F—

—G—

—H—

—I—

—L—

–T–

–U–

–V–

–W–

–Y–

Photo Credits

Page	Source
Cover	UPI
6	Elaine Fichera
7	Elaine Fichera
9	Camerique Stock Photos
10	Elaine Fichera
12	UPI
20	Oldsmobile Div. of General Motors
21	Captain Zoom
23	Sandy Kerns (left)
	Elaine Fichera (right)
26	Oldsmobile Div. of General Motors (top)
	Camerique Stock Photos (bottom)
27	Elaine Fichera
30	Elaine Fichera
33	Owen Franken - Stock, Boston
35	Elaine Fichera
39	Vannucci Foto-Services
41	H. Armstrong Roberts (top left)
	A. Devaney, Inc., N.Y. (bottom left)
	John A. DeCindis - Stock, Boston (top right)
	Peter Wandermark - Stock, Boston (bottom right)
45	Elaine Fichera
47	UPI (left)
	John Running - Stock, Boston (right)
49	Stock, Boston
51	Elaine Fichera
52	Sandy Kerns (top)
	Elaine Fichera (bottom)
53	Vannucci Foto-Services (bottom)
55	Elaine Fichera
57	UPI
58	UPI
59	Captain Zoom
60	H. Armstrong Roberts
61	H. Armstrong Roberts (top)
	© Bohdan Hrynewych - Stock, Boston (bottom)
65	A. Devaney, Inc., N.Y. (left)
	Elaine Fichera (right)
66	Elaine Fichera (top left, bottom left)
	UPI (right)
67	UPI
71	Owen Franken - Stock, Boston
75	Elaine Fichera (left)
	Christopher S. Johnson - Stock, Boston (right)
76	© Betty Medsger
78	Sandy Kerns
79	Captain Zoom
80	H. Armstrong Roberts (left)
	George Malave - Stock, Boston (right top)
	A. Devaney, Inc., N.Y. (right bottom)
81	Captain Zoom
83	UPI (left)
	Elaine Fichera (right)
86	A. Devaney, Inc., N.Y.
88	Elaine Fichera